SpringerBriefs in Environment, Security, Development and Peace

Volume 40

Series Editor

Hans Günter Brauch, Peace Research & European Security Studies, Mosbach, Germany

http://www.afes-press-books.de/html/SpringerBriefs_ESDP.htm
http://www.afes-press-books.de/html/SpringerBriefs_ESDP_40.htm

Ana Elizabeth Jardón Hernández ·
María Verónica Murguía Salas ·
Itzel Hernández Lara · Zoraida Ronzón Hernández

Multiple Discriminations

Perspectives on Social Inequality
of Vulnerable Groups in Mexico

Ana Elizabeth Jardón Hernández
Center for Research in Social Sciences
and Humanities
Autonomous University of the State
of Mexico (UAEMex)
Toluca, State of Mexico, Mexico

María Verónica Murguía Salas
Center for Research in Social Sciences
and Humanities
Autonomous University of the State
of Mexico (UAEMex)
Toluca, State of Mexico, Mexico

Itzel Hernández Lara
Faculty of Political and Social Science
Autonomous University of the State
of Mexico (UAEMex)
Toluca, State of Mexico, Mexico

Zoraida Ronzón Hernández
Center for Research in Social Sciences
and Humanities
Autonomous University of the State
of Mexico (UAEMex)
Toluca, State of Mexico, Mexico

ISSN 2193-3162   ISSN 2193-3170 (electronic)
SpringerBriefs in Environment, Security, Development and Peace
ISBN 978-3-031-85825-3   ISBN 978-3-031-85826-0 (eBook)
https://doi.org/10.1007/978-3-031-85826-0

© The Editor(s) (if applicable) and The Author(s), under exclusive license to Springer Nature Switzerland AG 2025

This work is subject to copyright. All rights are solely and exclusively licensed by the Publisher, whether the whole or part of the material is concerned, specifically the rights of translation, reprinting, reuse of illustrations, recitation, broadcasting, reproduction on microfilms or in any other physical way, and transmission or information storage and retrieval, electronic adaptation, computer software, or by similar or dissimilar methodology now known or hereafter developed.
The use of general descriptive names, registered names, trademarks, service marks, etc. in this publication does not imply, even in the absence of a specific statement, that such names are exempt from the relevant protective laws and regulations and therefore free for general use.
The publisher, the authors and the editors are safe to assume that the advice and information in this book are believed to be true and accurate at the date of publication. Neither the publisher nor the authors or the editors give a warranty, expressed or implied, with respect to the material contained herein or for any errors or omissions that may have been made. The publisher remains neutral with regard to jurisdictional claims in published maps and institutional affiliations.

This Springer imprint is published by the registered company Springer Nature Switzerland AG
The registered company address is: Gewerbestrasse 11, 6330 Cham, Switzerland

If disposing of this product, please recycle the paper.

Introduction

By identifying discrimination as a specific form of social inequality, this chapter presents the aim of the book: analyze the different manifestations of discrimination that occur within various vulnerable population groups, namely, international migrants, indigenous women, youth, and the elderly, with the goal of contributing to the construction of peaceful and inclusive societies in Mexico. This chapter introduces the context of the State of Mexico, where the practices of discrimination for each group analyzed take place, explains the research method that sustains the inquiry and present the sections that are part of the book.

This book is the result of the joint work of the Contemporary Social Processes research group at the Autonomous University of the State of Mexico. Based on the various individual research lines on international migration, indigenous labor mobility, youth and labor markets, old age, and age groups, it has been possible to find various points of convergence and common issues in the discussions and research results. In this regard, a prevailing element in the various research refers to the processes of discrimination that affect the exercise of rights of the populations focused on by this research team: international migrants, indigenous women, young people, and the aging.

Thus, the initiative arose to carry out an exercise focused on the analysis of discrimination processes in these groups and to put the research results into dialogue with a public interest issue and recognition of rights. Similarly, this effort is framed within the persistence of discrimination processes in Mexican society, as they are everyday practices that undermine the dignity of individuals and access to various rights. Their prevalence also contributes to the maintenance of social inequality processes based on prejudices and stigmas associated with physical characteristics, ethnic or national origin, sex, age, language, etc. It is a social problem that requires research and reflection to visualize and analyze its multiple manifestations in various contexts, as an essential step to propose public policy actions.

The international interest in eradicating discriminatory practices has been reflected in treaties such as the Sustainable Development Goals (SDGs), the Montevideo Consensus on Population and Development, as well as the International Labor Organization conventions on discrimination. These international instruments urge signatory States to take actions aimed at eradicating discriminatory practices, marking the institutional route to counteract them and promote access to justice, social inclusion, and dignified treatment.

In line with these international calls, discrimination has positioned itself as a topic on the public agenda of the Mexican state in the last two decades, and significant institutional efforts and normative changes have been made aimed at guaranteeing the right to non-discrimination. In that vein, the Mexican state carries out public actions at the national level but also in the federal entities that make up its territory.

Considering the latter, this book focuses on the analysis of discrimination processes, focusing on the State of Mexico, an entity whose relevance is not minor. It is the most populous federative entity nationwide with 16,992,418 inhabitants according to the 2020 Population and Housing Census (Instituto Nacional de Estadística y Geografía [INEGI], 2020), and it has an Economically Active Population (EAP) of 8,339,298 people as of the third quarter of 2023 (Gobierno de México, n.d.). Likewise, it is the second most important economy in the country, contributing 9% of the national Gross Domestic Product and being the third entity in attracting Foreign Direct Investment (Gobierno del Estado de México, n.d.).

In geographical terms, it is located in the central portion of the Mexican Republic and has a territorial extension of 22,499.95 km^2, and administratively, it is divided into 125 municipalities that have a wide heterogeneity in terms of climates and biodiversity, access to services and infrastructure, labor markets, etc. Another element to highlight is its border with Mexico City, which practically surrounds it to the north, east, and west (Fig. 1), which also implies an intense interstate dynamic in terms of labor, education, environment, and transportation, among others.

Likewise, the research of the working group has focused on various municipalities of the State of Mexico, including urban areas such as the municipality of Toluca (the capital of the entity) and Metepec, as well as metropolitan areas such as Tlalnepantla, Naucalpan, and Ecatepec. Research results focused on San Antonio la Isla and Tenancingo (to the south of the entity) are also included, as well as municipalities with rural communities and indigenous presence such as San José del Rincón, San Felipe del Progreso, Jocotitlán, and Atlacomulco (Fig. 1).

The research results integrated into this book account for the complexity of social contexts within the State of Mexico, while also highlighting the prevalence of discriminatory practices in different social spheres, with a particular emphasis on the labor sector: difficulties in accessing employment, labor exploitation, lack of recognition of their capabilities, absence of regulation and institutional support, etc., framed within various forms of inequality and power relations. As a result, an exercise focused on vulnerable population groups susceptible to discrimination in the State of Mexico has been proposed, within the framework of specific contexts and relationships, as explained below.

Regarding individuals in contexts of human mobility, the analysis focused on the State of Mexico requires a specific approach to the various dynamics of human mobility—emigration, immigration, transit, refugee status, and return—occurring in its territory, to visualize the complexity and dynamism that these processes have

Fig. 1 Geographic location of the State of Mexico and selected municipalities. *Source* Elaborated by Luz María Ledesma Reyes, 2024

acquired, particularly to recognize the situations, characteristics, and needs specific to each population. The study and analysis of these elements are inputs for proposing strategies aimed at the care and protection of this population.

The modalities prioritized in this book—refugee status, transit, and forced return—respond to their relevance in exploring the migration-discrimination dyad, as they are the flows that have gained a greater presence in the state, although not necessarily all of them have been the focus of attention in the public agenda of the state government. In this regard, the recent diagnosis presented by the Unit of Migration Policy, Registration, and Identity of Persons (UPMRIP, 2022) does not address the issue of refugee status. This absence contributes to the state of invisibility and lack of protection of individuals who have initiated their refugee status recognition application and/or who have obtained such recognition, as they seek to rebuild their lives in municipalities in the State of Mexico.

On the other hand, the UPMRIP diagnosis indicates that irregular transit migration represents one of the migratory flows that demands attention in the state, both due to its increased dynamics and the projections that support its continuity and rise. Among other factors, one of the explanatory factors for this dynamic is the transit routes from the State of Mexico to the northern regions of the country, involving municipalities such as Tepotzotlán, Toluca, and Chalco.

The third modality, forced return, associated in this work with the repatriation of nationals from the United States, is important because the State of Mexico is among the most affected entities in Mexico, ranking among the top five according to the origin of the population returned by US authorities. The composition and heterogeneity of these flows represent a challenge, while also demanding attention from an intersectional perspective, considering the greater presence of unaccompanied children and adolescents, as well as women, of various ages and with prolonged stays in the United States.

Regarding the indigenous population, data from the 2020 Population and Housing Census report 417,603 people aged three and over who speak an indigenous language in the State of Mexico, which corresponds to 2.6% of the state's population (INEGI, 2020). Five indigenous groups native to the State of Mexico are recognized: Nahua, Matlatzinca, Tlahuica, Otomi, and Mazahua, concentrated in rural areas but also participating in processes of labor mobility to urban areas in search of income for their households. Additionally, the state has been a destination for other ethnic groups such as immigrants from various states of the country. This is the case for Mixtecos, Zapotecs, or Mazatecos, with a greater concentration in urban areas adjacent to Mexico City (Consejo Estatal para el Desarrollo Integral de los Pueblos Indígenas del Estado de México [CEDIPIEM], 2023).

In addition to language use, indigenous communities are culturally distinguished by their forms of social organization, a common origin and a strong link with their native territory, and their various symbolic references such as clothing, which allow for their identity reproduction (Escalante, 2009). However, the indigenous population of Mexico has been characterized by its socioeconomic vulnerability, lag in access to health, education, justice, and employment (Consejo Nacional para Prevenir la Discriminación [CONAPRED], 2023; Téllez et al., 2013), and facing discrimination due to their ethnic origin because of stereotypes associating indigenous people with poverty, backwardness, and laziness (CONAPRED, 2023). This undermines their access to rights and dignified treatment, as well as the preservation of their cultures.

Indigenous peoples do not form a homogeneous group, and from an intersectional perspective, it has been recognized that indigenous women suffer greater discrimination due to their gender. Likewise, the discriminatory practices they face occur in various spheres and contexts. Therefore, this book focuses on women belonging to the Mazahua ethnic group who have migrated to Mexico City in search of work. Their labor integration in urban areas is characterized by labor precarity, particularly evident for those women working as domestic workers, a job distinguished by its lack of valuation and regulation, highlighting the intersection between ethnic origin, gender, and social class in the discrimination they face daily in the workplace.

On the other hand, market flexibility is a measure promoted by international organizations and exercised in the country to make the employer-employee relationship more volatile, reduce compensation for dismissals, evading direct responsibilities with employees through outsourcing, among other measures, aiming for transnational corporations to quickly adjust to international demands for offers, demands, and economic crises (Bensusán, 2010).

Within this context, significant gaps are created between population groups, where young people are considered among the most vulnerable and face greater obstacles to their full social inclusion (Weller, 2009). Consequently, young people having difficulty obtaining decent work leads to the underutilization of the labor force in this age group, high unemployment rates, higher proportions of informal employment, and an increase in precarious jobs.

According to Cruz et al. (2017), youth is a psychosocial stage where various transitions occur, influencing the course of life such as completing formal education, entering the labor market, establishing independent residence, first union, and birth to the first child, among others. Therefore, if young people begin their work trajectories with deficiencies, restrictions, and inequalities, it implies instability and uncertainty in other areas of life, both present and future.

Therefore, this book proposes to emphasize a group of young people who work and pursue their professional studies at the main public state university in the State of Mexico, as a way to approach the difficulties they face on the path to their full labor and, therefore, social inclusion. It is recognized that these limitations are expressions of discrimination since deficient remuneration, discontinuous work schedules, and social unprotectedness are characteristics of labor precariousness (Esparza, 2012), manifesting obstacles and restrictions to the full exercise of the fundamental rights of young people.

Considering age, it is possible to point out that Mexico is a country that is aging, where the population pyramid is inverted (narrowing the base and widening the top), as a result of what demography has called demographic transition, understood in the Mexican context as changes in fertility and mortality that occurred throughout the 20th century. In 1930, life expectancy was 34 years and by 1995 it had reached 73, a moment when the National Population Council (CONAPO) would construct projections that have drawn special attention to this point. If current demographic trends continue, where the life expectancy for a Mexican person is 76 years, it was estimated in 2012 that by 2020 the proportion of the population aged 60 and over would be 11.9 percent nationally, and indeed, in 2020, there is a 12% increase in older people compared to the total population. This means that the projections, which have been in place for more than a decade, remain valid, and it is expected that by 2050, life expectancy will be 83.6 years (Consejo Nacional de Población [CONAPO], 2019).

Thus, this book proposes to analyze old age and aging as conditions that lead people aged 60 and over to experience constant discrimination, which is relevant to account for the different ways in which their rights are violated in everyday life. While the analysis of rights violations usually starts from the observation of the State, recognizing that older people experience greater vulnerability than other age groups to diseases (physiological decline), poverty or impoverishment (reduction in income, retirement, or labor discrimination), and social marginalization (decrease in the flow of social relationships), these conditions are evident when in family and community dynamics, discriminatory acts toward older persons are legitimized, assumed to be socially accepted, and reproduced by the community, if not as a social norm, then as a consensus. Different forms of discrimination in old age in the State of Mexico are analyzed based on this premise.

To contribute and delve deeper into the understanding of this social issue in the State of Mexico, this work aims to analyze the different expressions of discrimination experienced by various groups of vulnerable populations, to recover experiences and perspectives from specific cases that highlight the urgency of strengthening strategies and actions aimed at building peaceful, tolerant, and inclusive societies in Mexico.

The theoretical framework of this book embraces a human rights perspective as a key resource to counteract discriminatory practices. Through this lens, the interpretation of the experiences recounted here is situated within the discourse of structural and intersectional discrimination, as the complexity of this issue can hardly be analyzed by dissociating the cultural, social, and political elements that give rise to it, nor can it be understood without recognizing the various variables of social differentiation that express multiple forms of discrimination and vulnerability based on characteristics associated with age, ethnic origin, sociocultural condition, nationality, or legal status in society.

The methodological approach adopted in this work begins with an exercise in descriptive statistics, serving as an analytical approach used to describe the dynamics of discrimination among the populations of these four vulnerable groups in Mexico. The description is based on data from the National Survey of Discrimination (ENADIS) 2017 and 2022, to conduct a comparative analysis of variables related to prejudices, denial of rights, and other aspects that highlight the multiple discriminations experienced by these populations.

The collection of information and interpretation of specific cases were conducted using qualitative methodology. Therefore, a phenomenological perspective is adopted, as it is the theoretical orientation that allows for the retrieval and interpretation of diverse experiences of discrimination from the standpoint and voice of the subjects (Taylor & Bogdan, 1994), as well as for a deeper exploration of the subjectivity of the informants who have been part of this research.

To this end, the use of the biographical method allowed for the collection of narratives and accounts aimed at understanding situations, contexts, and reasons associated with various processes of social inequality delimited according to the objectives of the projects carried out by this research team during the period 2020-2023.[1] The relevance of recovering this method also responds to the possibilities it offers to understand the specificities and differentiated positions that individuals

---

[1] The research projects that support the content of this work are as follows:

- Analysis of (re)employment processes of voluntarily and involuntarily returned migrant population from the United States in municipalities of the State of Mexico (Registration Key UAEMéx: 4975/2020CIF). Funding provided by UAEMéx, 2020. Responsible: Ana Elizabeth Jardón Hernández.
- Is Tijuana my limit? Social processes derived from the implementation of the Stay in Mexico program (Registration Key UAEMéx: 6315/2020CAP). Funding provided through the Program for Teacher Professional Development of the Ministry of Public Education, 2020. Responsible: Ana Elizabeth Jardón Hernández
- Work and gender inequalities in the context of labor mobility in rural communities in the northwest of the State of Mexico in the face of the COVID-19 pandemic (Registration Key UAEMéx: 6502/2022CIB). Funding provided by UAEM, 2022. Responsible: Itzel Hernández Lara.
- Being labor: work experiences of university students and their understandings of the meaning of work (Registration Key UAEMéx: 5112/2020SF). A project without funding. Responsible: María Verónica Murguía Salas.

construct based on their experiences and individual sensitivities. In other words, Pujadas (2000, p. 130) points out that:

> Biographical material helps dispel the specter of subject typification as representative or characteristic of a particular sociocultural order by introducing subjective and personal biases, which allow for the highlighting of different positions, sensitivities, and individual experiences.

The adoption of this method allows for building bridges with the intersectional perspective since the perspectives of the experiences lived by each individual are shaped by specific attributes and characteristics, which at certain times and spaces complicate their condition of vulnerability, lack of protection, and violation of their rights.

Among the approaches associated with the biographical method, the biographical narrative was used, aiming to make a literal record of the sessions with the interviewed individuals and their provided testimonies (Sanz, 2005). Therefore, as can be seen in the narratives of the interviewed individuals, the transcription of orality is pragmatic rather than grammatical.

Under the ethical considerations of all research, it is important to note that the presentation of the cases in this work has the written and/or verbal authorization of the interviewed individuals. To preserve anonymity, pseudonyms are used, so the names assigned to the collaborators of these research projects, in some cases, refer to any other name by which the person prefers to be called, and in others, they are assigned indiscriminately according to the person's gender.

From the Social Sciences and Humanities, this text seeks to contribute to the field of knowledge about discrimination processes in Mexico in two aspects. Firstly, by providing evidence and recognition of discrimination as a national problem, it offers an analysis focused on the State of Mexico, an entity that shows significant social complexity and about which it is necessary to build knowledge regarding discrimination processes as a necessary step to propose public action actions aimed at improving the living conditions of vulnerable groups, their social inclusion, and full access to rights. Secondly, this book carries out an analysis exercise from an intersectional approach, where the articulation of various discriminations in particular areas is evident. This allows for emphasis on power relations, prejudices, and the absence of legal and institutional instruments as elements that come into play in perpetuating discriminatory practices in areas such as work and everyday life toward vulnerable groups. Thus, it accounts for the complexity of discriminatory practices and the challenge of dismantling the structures that support them as a way to contribute to the eradication of social inequalities.

Finally, the structure of this work, in addition to the introduction, consists of seven chapters, of which the first sets out the theoretical and conceptual framework

---

- Care for the elderly in the Valley of Mexico. Sociocultural practices surrounding COVID-19 illness (Registration Key UAEM: 6797/2022CIB). Funding provided by UAEM, 2022. Responsible: Zoraida Ronzón Hernández.

from which to understand and analyze the processes and different forms of discrimination. This theoretical framework serves as the basis for reading the results of the contextual chapter that integrates the analysis of practices, perceptions, motives, and areas in which international migrants, indigenous women, youth, and older people are subjects of discrimination.

The following four chapters delve into the dynamics and specific experiences for each of these groups in the State of Mexico. The presentation of these cases aims to make visible the various spheres of discrimination in specific contexts, as well as the perceptions and actions of those who experience it. The concluding chapter presents a series of reflections and recommendations on the need to implement actions to guarantee equal treatment and the right not to be discriminated against.

References

Bensusán, G. (2010). Las reformas laborales y el corporativismo mexicano: alternativas en Europa y América Latina. En I. Bizberg (Ed.), *México en el espejo latinoamericano ¿Democracia o crisis?* (pp. 297–358). El Colegio de México, Fundación Konrad Adenauer.

Consejo Estatal para el Desarrollo Integral de los Pueblos Indígenas del Estado de México. (2023).

Consejo Nacional de Población. (2019). *Colección. Proyecciones de la población de México y las entidades federativas 2016–2050*. Secretaría de Gobernación. https://www.gob.mx/cms/upl oads/attachment/file/487366/33_RMEX.pdf

Consejo Nacional para Prevenir la Discriminación. (2023). *Discriminación en contra de la población y pueblos indígenas*. Ficha Temática. CONAPRED. http://www.conapred.org.mx/wp-content/uploads/2024/02/FT_Pindigenas_Noviembre2023.pdf

Cruz, R., Vargas, E., Hernández, A. y Rodríguez, O. (2017). Adolescentes que estudian y trabajan: factores sociodemográficos y contextuales. *Revista Mexicana de Sociología* 79(3), 571–604. https://revistamexicanadesociologia.unam.mx/index.php/rms/article/view/57679/51146

Escalante, Y. (2009). Derechos de los pueblos indígenas y discriminación étnica o racial. *Cuadernos de la igualdad*, (11). https://sindis.conapred.org.mx/investigaciones/derechos-de-los-pueblos-indigenas-y-discriminacion-etnica-o-racial/

Esparza, M. (2012). Empleo insuficiente y deterioro de las condiciones laborales en Zacatecas en los albores del nuevo siglo. *Paradigma económico. Revista de economía regional y sectorial*, 4(2), 61–84. https://paradigmaeconomico.uaemex.mx/article/view/4782/3187

Gobierno de México. (n.d.). *Data México. Estado de México. Economía*. Gobierno de México. https://www.economia.gob.mx/datamexico/es/profile/geo/mexico-em?redirect=true

Gobierno del Estado de México. (n.d.). *Indicadores económicos*. Secretaría de Desarrollo Económico. https://desarrolloeconomico.edomex.gob.mx/indicadores_economicos

Instituto Nacional de Estadística y Geografía. (2020). *Censo de población y vivienda 2020*. INEGI. https://www.inegi.org.mx/programas/ccpv/2020/#tabulados

Pujadas, J. (2000). El método biográfico y los géneros de la memoria. *Revista de Antropología Social*, 9, 127–158.

Sanz, A. (2005). El método biográfico en investigación social: potencialidades y limitaciones de las fuentes orales y los documentos orales. *Asclepio*, LVII, 99–115. https://doi.org/10.3989/asclepio.2005.v57.i1.32.

Taylor, S. y Bogdan, R. (1994). *Introducción a los métodos cualitativos de investigación* (2ª ed.). Paidós.

Téllez, Y., Ruiz, L., Velázquez, M. y López, J. (2013). Presencia indígena, marginación y condición de ubicación geográfica. *La situación demográfica de México 2013*.

Unidad de Política Migratoria, Registro e Identidad de Personas. (2022). *Diagnóstico de la movilidad humana en el Estado de México.* UPMRIP https://portales.segob.gob.mx/work/models/Politi caMigratoria/CPM/foros_regionales/estados/centro/info_diag_F_centro/diag_Edomex.pdf

Weller, J. (2009). El fomento de la inserción laboral de grupos vulnerables. Consideraciones a partir de cinco estudios de caso nacionales. CEPAL.

Contents

1	**The Concept of Discrimination, an Ongoing Discussion**	1
	1.1 The Construction and Complexity of the Discrimination Concept	2
	1.2 Social and Cultural Roots of Discrimination	4
	1.3 The Rights-Based Approach to Eradicate Discrimination	6
	1.4 Structural Discrimination	12
	1.5 Intersectional Discrimination	13
	1.5.1 Discrimination Based on Gender (Sex)	15
	1.5.2 Age Discrimination	16
	1.5.3 Discrimination Based on National Origin	16
	1.5.4 Discrimination Based on Migratory Status	17
	1.5.5 Discrimination Based on Ethnic Origin	17
	1.5.6 Labor Discrimination	18
	1.6 Conclusion	19
	References	20
2	**Current Context and Challenges Regarding Discrimination in Mexico**	23
	2.1 Discrimination in the Public Agenda in Mexico	23
	2.2 Discrimination Overview: Actors, Practices, and Denial of Rights	28
	2.3 Conclusion	40
	References	41
3	**"The Mexicos as Perceived and Lived": An Approach to the Discrimination Experiences of Populations in Human Mobility Contexts**	45
	3.1 Actions for Building Peaceful, Inclusive, and Non-discriminatory Societies	46

	3.2	Discrimination in the Experience of Mobility and Migration	49
		3.2.1 Escaping the Crisis: In Search of Refugee Status and Family Reunification	50
		3.2.2 Cultural Vulnerability Among Transient Migrant Populations	53
		3.2.3 From Stigma to Denial of Rights for Deported Population from the State of Mexico	56
	3.3	Conclusion	59
	References		61
4	**Discrimination Against Indigenous Women from the State of Mexico Who Work as Domestic Workers in Mexico City**		**63**
	4.1	Mazahua Migration to Mexico City and Labor Niches	65
	4.2	Brief Considerations on Domestic Work in Mexico	67
	4.3	Indigenous Women Working as Domestic Workers: Multiple Discriminations	68
	4.4	Public Actions to Dignify Domestic Work: Legal Framework and Cultural Change	72
	4.5	Conclusion	74
	References		74
5	**Labor Discrimination: The Case of Young People Who Study and Work**		**77**
	5.1	Study and Work: Fundamental Rights in the International Agenda	78
	5.2	Challenges and Dilemmas of Studying and Working in the Young Population	80
	5.3	Studying and Working: A Strategy Limited by the Labor Market	81
	5.4	Experiences of Young People Who Study and Work at the Autonomous University of the State of Mexico	84
	5.5	Conclusion	88
	References		89
6	**Older People as Subjects of Discrimination Due to a Lack of Oversight of their Rights**		**91**
	6.1	Older People on the Agenda for the Recognition of Their Rights	92
	6.2	Notes on Demographic Aging	93
	6.3	Ageism and Age Discrimination as "New" Forms of Discrimination	95
	6.4	From Discriminatory Experiences to the Legitimization of Violated Rights	96
	6.5	Ageism Through Intersectionality	101
	6.6	Conclusion	102
	References		103

7	**Conclusions and Recommendations**		105
	7.1 The Proposals		105
		7.1.1 International Migrant Persons	106
		7.1.2 Indigenous Women	107
		7.1.3 Young Population	108
		7.1.4 Older People	109
	7.2 The Contributions		110

On the Institution 113

On the Authors 115

Bibliography 119

Abbreviations

ACNUR	La Agencia de la ONU para los Refugiados
BBVA	Banco Bilbao Vizcaya Argentaria
CACEH	Centro Nacional para la Capacitación Profesional y Liderazgo de las Empleadas del Hogar A.C
	National Center for Professional Training and Leadership of Domestic Workers
CDI	Comisión Nacional para el Desarrollo de los Pueblos Indígenas
	National Commission for Indigenous People
CEDAW	Convención sobre la eliminación de todas las formas de discriminación contra la mujer
	Convention on the Elimination of All Forms of Discrimination against Women
CEDIPIEM	Consejo Estatal para el Desarrollo Integral de los Pueblos Indígenas del Estado de México State Council for the Integral Development of Indigenous Peoples in State of Mexico
CEJIL	Centro para la Justicia y el Derecho Internacional
	Center for Justice and International Law
CELADE	Centro Latinoamericano y Caribeño de Demografía
	Latin American and Caribbean Demographic Center
CEPAL	Comisión Económica para América Latina y el Caribe
CM	Consenso de Montevideo sobre población y desarrollo
COESPO	Consejo Estatal de Población—Estado de México
	State Population Council—State of Mexico
COMAR	Comisión Mexicana de Ayuda a Refugiados
	Mexican Commission for Refugee Assistance
CONAPO	Consejo Nacional de Población
	National Population Council
CONAPRED	Consejo Nacional para Prevenir la Discriminación
	National Council to Prevent Discrimination

COPRED	Consejo para Prevenir y Eliminar la Discriminación de la Ciudad de México
	Mexico City Council to Prevent and Eliminate Discrimination
DH	Derechos Humanos
ECLAC	Economic Commission for Latin America and the Caribbean
ENADIS	Encuesta Nacional de Discriminación
	National Survey of Discrimination
GCM	Global Compact for Safe, Orderly, and Regular Migration
	Pacto Mundial para la Migración Segura, Ordenada y Regular
HIV/AIDS	Human Immunodeficiency Virus/Acquired Immunodeficiency Syndrome
HR	Human Rights
ILO	International Labour Organization
IMJUVE	Instituto Mexicano de la Juventud
	Mexican Institute of Youth
IMSS	Instituto Mexicano del Seguro Social
	Mexican Social Security Institute
INEGI	Instituto Nacional de Estadística y Geografía
	National Institute of Statistics and Geography
INM	Instituto Nacional de Migración
	National Institute of Migration
IOM	International Organization for Migration
ISSEMyM	Instituto de Seguridad Social del Estado de México y sus Municipios Social Security Institute of the State of Mexico and Municipalities
LFPED	Ley Federal para Prevenir y Eliminar la Discriminación
	Federal Law to Prevent and Eliminate Discrimination
LGBTIQA+	Lesbiana, Gay, Bisexual, Trans, Intersexual, Queer, Asexual/Agénero y otras diversas orientaciones e identidades de género
	Lesbian, Gay, Bisexual, Trans, Intersex, Queer, Asexual/Agender and other diverse gender orientations and identities
MC	Montevideo Consensus on Population and Development
MDGs	Millennium Development Goals
NU	Naciones Unidas
OAS	Organization of American States
ODM	Objetivos de Desarrollo del Milenio
ODS	Objetivos de Desarrollo Sostenible
OEA	Organización de los Estados Americanos
OHCHR	Oficina del Alto Comisionado de las Naciones Unidas para los Derechos Humanos
	The Office of the High Commissioner for Human Rights
OIM	Organización Internacional para las Migraciones
OIT	Organización Internacional del Trabajo
OJI	Observatorio de la Juventud en Iberoamérica
	Youth Observatory in Latin America

OMS	Organización Mundial de la Salud
OPS	Organización Panamericana de la Salud
OXFAM	Comité de Oxford de Ayuda contra el Hambre
	Oxford Committee for Famine Relief
PAHO	Pan American Health Organization
PEA	Población económicamente activa
	Economically Active Population
PRONAIND	Programa Nacional para la Igualdad y No Discriminación
	Program for Equality and Non-Discrimination
RAE	Real Academia Española
	Royal Spanish Academy
SDGs	Sustainable Development Goals
UAEMéx	Universidad Autónoma del Estado de México
	Autonomous University of the State of Mexico
UN	United Nations
UNESCO	Organización de las Naciones Unidas para la Educación, la Ciencia y la Cultura
	United Nations Educational, Scientific and Cultural Organization
UNHCR	The UN Refugee Agency
UPMRIP	Unidad de Política Migratoria, Registro e Identidad de Personas
	Unit of Migration Policy, Registration, and Identity of Persons
VIH/SIDA	Virus de Inmunodeficiencia Humana/Síndrome de inmunodeficiencia Adquirida
WHO	World Health Organization

List of Figures

Fig. 1.1	Intersectional analysis framework for discrimination. *Source* Own elaboration based on the proposal by Guzmán and Jiménez (2015, p. 605)	15
Fig. 2.1	Consultations and guidance provided by CONAPRED, 2010–2022. *Source* Own elaboration based on the annual reports of CONAPRED (n.d.a, 2022, 2023)	26
Fig. 2.2	Perception of the population belonging to vulnerable groups regarding the respect for their rights, 2017–2022. *Source* Own elaboration based on ENADIS (2017, 2022)	30
Fig. 2.3	Main issues considered by population group, 2022. *Source* Own elaboration based on ENADIS (2022)	32
Fig. 2.4	Population agreeing with prejudices attributed to their population group, 2022. *Source* Own elaboration based on ENADIS (2022)	33
Fig. 2.5	Denial of rights by population group and gender in, the last five years 2017 and 2022. *Source* Own elaboration based on ENADIS (2017, 2022)	35
Fig. 2.6	Persistence of discrimination by population group and gender in, the last five years 2017 and 2022. *Source* Own elaboration based on ENADIS (2017, 2022)	36
Fig. 2.7	Discrimination situations experienced by population group, last five years 2022. *Source* Own elaboration based on ENADIS (2022)	37
Fig. 2.8	Reasons for discrimination by population group, last twelve months 2022. *Source* Own elaboration based on ENADIS (2022)	38
Fig. 2.9	Areas of discrimination by population group, last twelve months 2022. *Source* Own elaboration based on ENADIS (2022)	40
Fig. 3.1	Vulnerability of the migrant population. *Source* Own elaboration based on Bustamante (2002, p. 341)	54

Fig. 3.2	Stance on the Mexican government's actions regarding the Central American population. *Source* Own elaboration based on ENADIS, 2022	56
Fig. 3.3	Attitudes of rejection and acceptance towards Central American Population by State. *Source* Own elaboration based on ENADIS, 2022	57
Fig. 4.1	Municipalities of origin of Mazahua migration to Mexico City. *Source* Elaborated by Luz María Ledesma Reyes based on Hernández, 2025	66
Fig. 5.1	Pathways of young people who study and work. *Source* Own elaboration based on 25 interviews conducted with young people who study at UAEMéx and work, 2020–2021	84
Fig. 6.1	Ramifications of demographic transition. *Source* Own elaboration based on the analysis of demographic transition	95
Fig. 6.2	Negative characteristics linked to aging. *Source* Own elaboration	96
Fig. 6.3	Examples of intersectional discrimination for a man and a woman. *Source* Own elaboration based on testimonies from older persons	102
Fig. 7.1	Elimination of violence from a rights perspective. *Source* Own elaboration	106

List of Tables

Table 1.1	Characteristics of structural discrimination	13
Table 2.1	Main causes of discrimination in complaints handled by CONAPRED, 2017–2022	27
Table 3.1	GCM objectives with the highest number of programs and/or actions, 2022	48
Table 6.1	Population aged 60 and over as a percentage of the total population in Mexico	94
Table 6.2	Population aged 60 and over as a percentage of the total population in State Mexico	94

Chapter 1
The Concept of Discrimination, an Ongoing Discussion

This section of the book aims to position and support the theoretical and conceptual perspective that underpins the subsequent sections, arguing that discrimination is sustained and reproduced through power relations, evident in the domination-oppression dyad. Thus, the pursuit of eliminating discrimination must be based on a rights-based approach. While the analysis of discrimination may involve a multitude of variables that can be disparate, a human rights perspective is vital for achieving peaceful and equitable societies.

To gain a comprehensive understanding of the complexity of the concept of discrimination, this section is divided into five parts. The first section highlights that discrimination is not only the inferior treatment of a person or group for unjustified reasons but also emphasizes that such treatment leads to the annulment or restriction of the enjoyment of human rights. The second section addresses two positions on the social and cultural roots of discrimination that can give meaning to discriminatory expressions in society, as well as ways to prevent or eradicate them. Building on this, the third section explains how efforts are made to contribute to the achievement of the SDGs and the Montevideo Consensus, which serve not only as guides for analyzing reality but also as milestones to align and propose public policies and take action against discriminatory expressions.

The final sections focus on two types of discrimination: structural and intersectional. Both approaches provide guidelines for understanding and analyzing four groups of people who are subject to various forms of discrimination and will be specifically addressed in this book: international migrants, indigenous women, youth, and older persons.

© The Author(s), under exclusive license to Springer Nature Switzerland AG 2025
A. E. Jardón Hernández et al., *Multiple Discriminations*,
SpringerBriefs in Environment, Security, Development and Peace,
https://doi.org/10.1007/978-3-031-85826-0_1

1.1 The Construction and Complexity of the Discrimination Concept

The term "discriminate" refers to the selection or differentiation of one element from another. Rodríguez (2023) refers to this meaning as lexicographic, where there is no negative or derogatory sense, as it simply denotes the action of separating, distinguishing, or choosing. However, the focus of this text is on the second common definition of discrimination, which is to "treat a person or group with inferiority based on racial, religious, political, etc., motives" (Real Academia Española [RAE] 2023). Nonetheless, to address how certain population groups are subject to different forms of discrimination, there is a need to discuss the concept in terms of its components, uses, motives, types, and analytical perspectives.

First and foremost, a central element of the definition of discrimination is treatment, which is understood as how one interacts or engages in a relationship with one or more individuals. In this case, emphasis is placed on the negative sense, implying an asymmetrical relationship where one person is presumed to be superior over another considered inferior. The treatment of inferiority or differentiation is based on preconceived criteria such as gender, age, place of origin, religious beliefs, and physical appearance, among others, which are considered valid justifications for actions that undermine human dignity.

The visibility of discrimination against certain groups over others leads to the acceptance of the very domination of some over others. This, to a large extent, has been extrapolated in contemporary societies due to the exacerbation of capitalism.[1] As Godelier (1986) argued, the domination-oppression relationships have been as ancient as humanity itself. Thus, in contemporary capitalist societies, discrimination turns against the dominated as a means to reinforce power through hegemony.

The form of domination between sexes observed by Godelier (1986) in pre-industrial societies, where women's rights were granted based on men's recognition, was evidence of women's subordination to men. Additionally, violence was part of this mechanism of male-female dominance. Godelier suggests that this type of domination-subordination relationship implies a "certain consent" from the dominated group, stemming from "the existence of social and psychological mechanisms to create this consent" (p. 45). Such domination also relies on other "mechanisms" (in Godelier's terms) of culture that produce, reproduce, and replicate hegemonic ideologies of legitimization, almost by antonomasia, of some over others.

In this line of thought, the idea of hegemony is a concept derived from nineteenth century Marxist theory, which, according to Giacaglia (2002), "emerges as a response to a crisis that questions traditional Marxist categories to explain contingency" (p. 151). The author explains that reformulations of Marxism regarding hegemony as a theoretical concept must have a standpoint from which to

---

[1] The discussion and analysis of capitalism have been widely observed from various areas of knowledge. For this work, it only serves as a reference within the context of Godelier's works (1986).

discuss a "new" hegemony, although Godelier (1986) would likely argue that hegemonies are hardly ever truly new; it is only the social actors that change. What must be observed when analyzing hegemony are the historical-social conditions in which that hegemony appears in a specific socio-cultural space.

Thus, discussions regarding hegemony and the domination-subordination relationship have been ongoing across various disciplines, never disappearing, as conditions of inequality, inequity, and poverty persist, prompting continued observation, analysis, and discussion to seek alternatives leading to more egalitarian societies.

In the same vein, Gramsci (1975) discusses subalternity, or subaltern groups, to refer to groups subjected in this dynamic of dominated and dominators. Being aware of this subalternity, these groups can emerge from domination through autonomy. Therefore, in contemporary societies, the struggle for the respect of all individuals' rights and consequently, the reduction of inequality/domination, would lead to the exercise of group and individual autonomy.

Thus, the treatment of inferiority is a central characteristic of the definition of discrimination provided by the RAE. However, it lacks the repercussions that result from such treatment. Specifically, it refers to the harm that can translate into the annulment or restriction of the enjoyment of human rights (Comisión Nacional de los Derechos Humanos [CNDH] 2012). In line with this, it is by Rodríguez (2023), points out that "the relevant semantic field to establish a theoretically correct and politically and legally productive definition of *discrimination* is none other than that of human or fundamental rights" (p. 36).[2] With this recognition, it's understood that discrimination isn't just about making a distinction and taking actions detrimental to an individual or group of people, but it constitutes a violation of human rights. This perspective is useful for analyzing cases concerning international migrants, indigenous women, youth, and older persons, as it exposes situations and experiences that highlight the denial of rights and the contexts in which they manifest.

The focus on human rights is not recent, as the United Nations General Assembly proclaimed the Universal Declaration of Human Rights in 1948. Article 7 emphasizes access to the rights stipulated therein and the prohibition of discrimination: "All are equal before the law and are entitled without any discrimination to equal protection of the law. All are entitled to equal protection against any discrimination in violation of this Declaration and against any incitement to such discrimination" (Naciones Unidas [NU] n.d.). Under this approach, discrimination should be understood as the denial of fundamental rights for unjustifiable reasons such as age, sex, religion, place of origin, among others. Therefore, actions must be taken to ensure the "right to have rights" (Rodríguez 2023, p. 43). With this perspective, this book aims to emphasize the recognition of the rights of various groups that have been subject to violations of their guarantees through discriminatory acts, contributing to the construction of peaceful and inclusive societies.

---

[2] The word "discrimination" in italics is from the original text.

1.2 Social and Cultural Roots of Discrimination

One way to understand discriminatory acts in our societies is through what Cortina (2017) calls "aporophobia." This term is coined by the author from the Greek word "*áporos*", which refers to the poor or the resourceless, thus "aporophobia" denotes the rejection of the poor. "It is urgent to name the rejection of the poor, the helpless because that attitude has a force in social life that is even greater precisely because it acts from anonymity" (Cortina 2017, p. 24).

According to Vargas et al. (2020), aporophobia serves to indicate that there are nuances in the rejection of what is different; it is not disdain for the different *per se* but rather due to "the condition of causing problems stemming from being poor" (p. 457). From this perspective, the authors mention that various attitudes of rejection such as xenophobia, misogyny, Christianophobia, Islamophobia, or homophobia all have a common root in the disdain for the poor, that is, aporophobia.

The rejection of the poor is not solely due to their economically and socially disadvantaged condition. The underlying reason for the disdain of the poor lies in the function they have in society. Cortina (2017) indicates that there exists a world of giving and receiving based on the political, economic, and social contract, where those who "seem to have something interesting to return as a reward" can enter (p. 6). The poor do not seem to have anything to contribute; rather, the author mentions that there is a perception that the poor will only bring problems, hence they are excluded from this world.

The poor threaten the well-being of certain individuals because in the exchange process, there is no reciprocity, and therefore, no gains. They "seem to contribute nothing positive to one's own survival and well-being" (Cortina 2017, p. 54). This calculative approach between giving and receiving puts the poor at a disadvantage, as they are attributed a negative connotation within the exchange process due to the perceived problems associated with their inclusion. Thus, Cortina (2017) points out that poverty introduces discrimination among people, as only a portion of humanity has sufficient means to organize their own lives and have a valuable position in the exchange process.

From this perspective, poverty is defined as the lack of necessary means for survival (Vargas et al. 2020) and serves as the driver of discrimination by threatening the well-being of others. In response, Cortina reflects on how discrimination can be avoided. It is not through giving more resources to the poor to increase their value, as one might assume. The change lies in two key elements of an ethics of cordial reason: "equal dignity of individuals and compassion, understood as the ability to perceive the suffering of others and to commit to preventing it" (Cortina 2017, p. 7).

Mutual recognition of dignity and compassion are considered the channels to overcome aporophobia and, consequently, discrimination. For Cortina (2017), change must occur through both formal and informal education, within the family, schools, through media, and in all aspects of public life. Additionally, the author emphasizes the need to build institutions and political, educational, and cultural

1.2 Social and Cultural Roots of Discrimination

organizations that aim in the same direction. In this sense, education would include, or should include, basic knowledge of the human rights that all people possess.

Up to this point, the term aporophobia is considered a novel approach to denote the underlying cause of various discriminatory expressions in our society: the rejection of the poor, the helpless. However, it is considered relevant to present another approach that explains the foundations of unequal treatment leading to the violation or limitation of human rights. To this end, the proposal of Rodríguez (2023) is recovered, who points out that there are two cultural drivers of discrimination: stigma and prejudice.

Similar to Rodríguez (2023), reference is made to the work of Irving Goffman to define and understand the term stigma. Initially, it is indicated that a stigma is a trait or attribute possessed by an individual that results in discredit from others due to a discrepancy between virtual and real social identity (Goffman 2006). He highlights three types of stigmas: (a) physical deformities, (b) individual character defects such as drug addiction or extremist political behavior, and (c) trivial stigmas such as nationality or religion. In all types of stigma, the same sociological traits are found, characterized by eliciting repulsion towards a person for possessing a trait that can forcibly draw attention away from their other attributes, thus "we believe, by definition, of course, that the person with a stigma is not fully human. Relying on this assumption, we practice various types of discrimination, through which, in practice, though often without thinking, we reduce their life chances" (Goffman 2006, p. 15).

Based on the above, Rodríguez indicates that stigma "is a cultural syntax of asymmetric classification of dominance and exclusion" (2023, p. 49). Following the author's argument, attributes can be considered negative in a relationship of subordination and dominance, having a disadvantageous effect on the stigmatized individual and an arbitrary power for the one assigning the stigma. Thus, stigma can be understood as the means that legitimizes differences, especially those that show distortion or fault, to maintain asymmetric relationships in the power system and carry out actions conducive to this, such as discriminatory practices.

However, Rodríguez points out that the process of stigmatization is possible through another cultural driver of discrimination: prejudice. This term denotes "an attitude of aversion or hostility towards a person belonging to a group, simply because they belong to that group, and it is consequently presumed that they possess the qualities ascribed to the group" (Allport 1954 in Rodríguez 2023, p. 52). It is through prejudice that a person is inscribed in a group and identified by generic characteristics, diluting particular elements.

Thus, prejudices disadvantage specific social groups, as they are generated and justified based on illegitimate benefits, pleasure, or satisfaction enjoyed by those who hold privileges (Rodríguez 2023). For example, the risk of discrimination against a young person is not only due to their particular traits but also because the adult group attributes negative characteristics to them inherent to their association with the youth group, which are justifiable in the eyes of adults, known as adultcentrism. Through this example, it can be understood that discrimination is directed at an individual for symbolically representing the characteristics of a specific social group, and such action finds its justification in the prejudices held by hegemonic groups.

However, it is important to note that "the likelihood of prejudice 'turning into action' is delimited by legal circumstances and the ethical and symbolic stimuli or constraints of the social group" (Rodríguez 2023, p. 59). In other words, the existence of prejudiced attitudes does not automatically lead to discriminatory actions. Prejudices about a social group may exist, but they may not necessarily lead to discrimination, as it depends on the social interactions between groups and the effectiveness of the regulatory framework that ensures access to fundamental rights for all individuals.

In this way, focusing on stigmas and prejudices is considered essential to visualize and analyze the interaction processes between individuals and, thereby, the systems of intersubjective relations that generate asymmetries of domination and exclusion. Therefore, it is understood that the cultural drivers of discrimination are framed within specific historical contexts and processes that result from a complex interweaving of political, economic, social, and ideological dimensions, among others.

According to Rodríguez, "stigmas and prejudices are at the root of the systematically despised behaviors suffered by different excluded or discriminated groups" (2023, p. 59). Due to this importance, the author proposes a definition of discrimination incorporating cultural drivers, to emphasize the social construction of unequal treatment and, in turn, generating mechanisms such as political action or education that influence at the level of shared symbolic order:

> Discrimination is a culturally rooted and socially pervasive behavior of contempt towards an individual or group of individuals based on prejudices or stigmas related to an undeserved disadvantage. Its effect, whether intentional or not, is to nullify or limit both their fundamental rights and freedoms and their access to socially relevant opportunities within their social context. (Rodríguez 2023, p. 60)

This definition highlights the symbolic and cultural foundations of discrimination, as well as the negative effects of differential treatment that hinder full access to human rights. Both elements are considered relevant to indicate that discriminatory acts towards a specific social group are not spontaneous; rather, they stem from historically entrenched social structures and are validated within the framework of reference in which realities are constructed. Therefore, the structural nature of discrimination becomes evident.

1.3 The Rights-Based Approach to Eradicate Discrimination

Social Sciences and Humanities have played a decisive role in highlighting how different population groups have been victims of positions of subordination that put them at a disadvantage and, consequently, where they are constantly vulnerable. This is the case of international migrants, indigenous women, youth, and the ageing, who are subject to different forms of discrimination, not only in specific aspects but also in everyday life, in activities so subtle that, at times, go unnoticed.

1.3 The Rights-Based Approach to Eradicate Discrimination

In this sense, since the last two decades of the twentieth century, contemporary society has been aware of the current situation, revealing the need to act not only as independent societies but as a global community. Hence, the new millennium arrived with a series of proposals and initiatives aimed at combating and overcoming poverty, which has been considered one of the most important variables in any society for discrimination, as it puts those in poverty in a situation of disadvantage/vulnerability.

The year 2000 echoed consciences and good intentions, expressed in what was known as the Millennium Development Goals (MDGs), whose main goal was the elimination of poverty, based on 8 objectives that together would achieve its elimination. These were: to reduce extreme poverty, achieve universal primary education, reduce child mortality rates, improve maternal health, combat epidemics of diseases such as HIV/AIDS, ensure environmental sustainability, and promote a global partnership for development. The actions of these MDGs apparently had an impact, as in 2015 extreme poverty reached its lowest known point (10%), which was considered a significant achievement given that in the early 1990s it reached almost 36% (Banco Mundial 2018).

Thus, the MDGs gave way, in 2015, to a broader version of themselves, the 2030 Agenda for Sustainable Development, with 17 Goals (SDGs 2030) and 169 targets, which have as their primary intention "to eradicate poverty, protect the planet and ensure prosperity for all as part of a new sustainable development agenda" (NU 2018), but also establishes that the 193 signatory countries must unite efforts to achieve them.

SDG number one, and the cornerstone of the others, is to end poverty in all its forms everywhere, and it maintains that:

> Poverty goes beyond the lack of income and resources to ensure sustainable livelihoods. Among its manifestations are hunger and malnutrition, limited access to education and other basic services, social discrimination and exclusion, and lack of participation in decision-making. Economic growth must be inclusive to create sustainable jobs and promote equality. (NU 2018, p. 7)

Therefore, the SDGs for the 2030 agenda become an international action guide in which a new generation of public policies must align to contribute, first and foremost, to the eradication of poverty and, as a logical consequence, to the disappearance of social inequalities in Latin America and the world, thus building a different world based on sustainable and affordable development, which, as challenging as it may seem, within a 15-year timeframe.

SDG number 10, Reducing Inequalities, aims to reduce inequality within and among countries. Despite a considerable decrease in poverty, "inequalities and significant disparities in access to health and education services and other productive assets still exist" (NU 2018, p. 47), making it clear that the economic growth a society may experience is not synonymous with creating an equitable society. Therefore, the three dimensions of sustainable development -economic, social, and environmental- must be taken into account. Thus, to reduce inequality it has been recommended the implementation of policies that aim to pay attention to the needs of disadvantaged and marginalized populations. For this purpose, target

10.3 aims to "guarantee equal opportunities and reduce inequality of outcomes, including by eliminating discriminatory laws, policies, and practices and promoting appropriate legislation, policies, and measures in that regard" (p. 48).

And has as an indicator: 10.3.1 is the "proportion of the population that has experienced personal discrimination or harassment in the last 12 months based on discrimination prohibited by international human rights law" (p. 48). Here, the collection of first-hand information in different states based on people's perceptions is taken into account.

Furthermore, but in the same vein, SDG 11 Sustainable Cities and Communities, sets out for 2030, "to make cities and human settlements inclusive, safe, resilient, and sustainable" (p. 51), and as difficult as it may seem to imagine this being possible or viable in 15 years, two goals are proposed that are crucial for this work: 11.1 to "ensure access for all to adequate, safe, and affordable housing and basic services and upgrade slums" by 2030, and 11.2 which aims to "provide access to safe, affordable, accessible, and sustainable transport systems for all and improve road safety, particularly through expanding public transport, with special attention to the needs of people in vulnerable situations, women, children, persons with disabilities, and older persons" (p. 51). Both goals demonstrate the urgent need to find efficient and tangible ways for cities to be spaces of inclusion and equity.

In the same vein, SDG 16 Peace, Justice, and Strong Institutions, is the objective that could directly contribute to making 10 and 11 not only possible but also viable and applicable, as it states: "promote peaceful and inclusive societies for sustainable development, provide access to justice for all, and build effective, accountable, and inclusive institutions at all levels" (p. 71), which in our country has involved a change in the criminal justice system, seeking to comply with goal 16.3 which aims to "promote the rule of law at the national and international levels and ensure equal access to justice for all" (p. 72), as well as goal 16.b which proposes "promote and implement non-discriminatory laws and policies for sustainable development" (p. 74), which would make any society inclusive, ensuring that every person, regardless of their age group, ethnic origin, or legal status in a country, deserves dignified treatment, access to justice, and full enjoyment of rights.

In this line of thought, but in a more regional order, the Montevideo Consensus (MC) is the document resulting from the search for overcoming inequalities for the achievement of sustainable development in Latin America in 2013 and had as antecedents the Latin American and Caribbean Consensus on Population and Development, in 1993, and the Regional Plan of Action for Latin America and the Caribbean on Population and Development in 1994, but with the particularity of emphasizing human well-being and dignity, in addition to sustainability (Comisión Económica para América Latina y el Caribe [CEPAL] 2013) and highlights:

> The universality, equality, transversality, comprehensiveness, inclusion, solidarity, equity, and dignity, as well as human rights in the application of approaches to all groups in vulnerable conditions, as well as related topics related to health, education, community, governance, and sustainability (p. 5).

1.3 The Rights-Based Approach to Eradicate Discrimination

It is a document that contemplates nine priority measures in population and development, with an agreement to achieve them, as well as lines of action, goals, and tentative indicators for each one, and it is based on three main themes: development, which involves eradicating poverty and breaking the cycles of exclusion and inequality; development focused on Human Rights (HR); and sustainable development (CEPAL 2015), under the premise that: "poverty in all its manifestations represents in itself the denial of rights, and its eradication is a moral imperative for the region that governments must assume" (CEPAL 2015, p. 6).

In this way, the MC seeks to break down different barriers that result in discrimination and, consequently, in inequalities and inequities among different population groups, so that from Priority Measure A, the "full integration of the population and its dynamics in sustainable development with equality and respect for human rights" is sought (CEPAL 2013, p. 7), highlighting the necessary application of HR so that, and only then, more just and inclusive societies are built, for which one of the agreements was "to apply a human rights approach with a gender and intercultural perspective in addressing population and development issues, and to increase efforts aimed at their recognition, promotion, and protection, to eliminate inequalities and promote social inclusion" (p. 7).

In the logic of the document and in addressing different population groups, the first one specifically mentioned is in Priority Measure "B. Rights, needs, responsibilities, and demands of children, adolescents, and youth" (CEPAL 2013, p. 8), as it not only recognizes this sector as rights holders but also highlights the need to address their needs. At the time of the document's drafting, one in four people in the region was aged between 15 and 29, making the demographic dividend seen as a "unique opportunity for social investment in adolescence and youth, based on intergenerational solidarity, an essential investment for sustainable development in its three pillars: social, economic, and environmental" (CEPAL 2013, p. 8). This led to Agreement Number 7 seeking to:

> Ensuring children, adolescents, and youth, without any form of discrimination, the opportunities to have a life free from poverty and violence, the protection and exercise of human rights, the availability of options, and access to health, education, and social protection. (CEPAL 2013, p. 8)

The priority measure C. Aging, social protection, and socioeconomic challenges, not only recognizes older people as subjects of rights but also states that countries must consider them as a key sector for the development of public policies. It acknowledges that "older people, due to their age and vulnerability, continue to be discriminated against and are victims of abuse and mistreatment, which consequently affects the enjoyment and exercise of their rights" (CEPAL 2013, p. 10). It also highlights the urgent need to address this sector given the demographic changes occurring in the region, as well as the epidemiological changes affecting the quality of life and society as a whole.

Based on the above, some of the most important agreements include:

> Develop policies at all levels (national, federal, and local) aimed at ensuring the quality of life, the development of potentials, and the full participation of older people, addressing the needs

for intellectual, emotional, and physical stimuli, and considering the different situations of men and women, with particular emphasis on the most vulnerable groups to discrimination (older people with disabilities, those lacking economic resources and/or pension coverage, and older people living alone and/or lacking support networks);

Eradicate the multiple forms of discrimination affecting older people, including all forms of violence against older women and men, taking into account the obligations of states regarding aging with dignity and rights. (CEPAL 2013, p. 11)

The significance of these agreements lies not only in the perspective of respecting and monitoring human rights but also in the consideration of building public policies in the absence of these and adopting an intersectional approach. Within this same framework lies Priority Measure F: International migration and the protection of the human rights of all migrants, where a:

About the evident and systematic violation of human rights suffered by migrant individuals due to racism, xenophobia, and homophobia, as well as the lack of guarantees of due process, and the specific issues affecting different groups in terms of discrimination, abuse, human trafficking, exploitation, and violence, particularly women, girls, boys, and adolescents. (CEPAL 2013, p. 18)

The unequal development experienced in the region generates discrimination that leads to the vulnerability of migrant individuals, resulting in precarious conditions in jobs, especially when individuals come from less developed economies or societies with greater poverty than the one they migrate to.

In response to this situation, one of the highlighted agreements is to "provide assistance and protection to migrant individuals, regardless of their migratory status, especially to those groups in vulnerable conditions" (CEPAL 2013, p. 18). Additionally, the commitment to comply with the International Convention on the Protection of the Rights of All Migrant Workers and Members of Their Families and the Vienna Convention on Consular Relations is reiterated, emphasizing that attention should not only be focused on destination countries but also on transit countries. Furthermore, it is agreed that the criminalization of migration should be avoided, while ensuring access to basic education and health services regardless of migratory status.

Measure H. Indigenous peoples: interculturality and rights emphasizes that indigenous peoples have all internationally recognized rights and should enjoy them without discrimination. It recognizes the cultural diversity in Latin America but also acknowledges the historical inequality that has affected indigenous peoples, resulting from structural poverty, leading to "higher levels of material poverty, exclusion, and marginalization, as well as less participation in decision-making in power structures and in bodies of popular representation, resulting in a violation of their human rights" (CEPAL 2013, p. 21). Hence, there is a clear need for "greater protection for their development, for the forms of violence suffered by indigenous children, adolescents, and young people, indigenous women, and indigenous older persons" (CEPAL 2013, p. 21). This underscores how different conditions or characteristics of individuals make certain population groups more vulnerable, and the agreements are aimed not only at respecting and guaranteeing their rights but also at restoring them if they have been previously violated, to "ensure that indigenous women, children, adolescents, and young people enjoy full protection and guarantees against all forms of violence and discrimination" (CEPAL 2013, p. 22).

In this way, the SDGs and the Montevideo Consensus serve as tools for analyzing and monitoring human rights, as well as for developing public policies that, under the observation of multiple discriminations, contribute to understanding how different groups or entities can influence the enjoyment of rights and, consequently, opportunities that can contribute to reducing or widening existing inequality gaps in society, and therefore, in development (Rodríguez 2019). It is also a tool that was built by the feminist movement in the seventies with the intention of making visible the different forms of discrimination that a woman can experience.

Thus, under the intersectionality approach, an individual can suffer discrimination because of being an older adult, being a woman, belonging to an indigenous ethnicity, living with a disability, or being in poverty. Each of these conditions contributes to placing them in a position of greater vulnerability. When faced with this set of circumstances, it is also called multiple discrimination, a term used in 2001 at the World Conference against Racism, Racial Discrimination, Xenophobia, and Related Intolerance (NU 2001).

Despite international efforts to establish guidelines that promote the defense of human rights and prevent discrimination, there are obstacles that limit their progress today. One of the main problems lies in the attribution of responsibilities and, therefore, in mechanisms to prevent or limit actions by those who perpetrate discriminatory acts. To expose the heterogeneity and complexity of the attribution of responsibilities, Rodriguez's proposal (2023) is revisited, who distinguishes two forms of the phenomenology of discrimination: direct and indirect. In the first case, the author indicates that it is relative to specific acts of discrimination, where one person causes a restriction or harm to the rights of another person. In this modality, discriminatory acts are individual, so the direct causal relationship between the individuals involved and the effects can be identified, resulting in detection of individual responsibilities.

However, there is indirect discrimination, which "relates to the historical formation of norms, patterns, and social institutions that, without being directed against a specific person [...] have the structural effect of maintaining and deepening the disadvantage of the group" (Rodríguez 2023, p. 61). As can be seen, responsibilities are diffuse because there are no defined figures between those who commit discriminatory acts or those who suffer from them. The discriminatory act cannot be precisely determined either since, occurring over a long period and being anchored in social norms, endorsed by institutions, it becomes naturalized in the social construct, making its detection difficult no matter how serious it may be, resulting in structural discrimination.

1.4 Structural Discrimination

According to Solís (2017), when it is stated that discrimination is a structural phenomenon, it is due to the existence of three traits: (a) it is based on a social order that is independent of individual wills; (b) it is a process of accumulation of disadvantages throughout the course of life and intergenerationally; and (c) it has macro-social consequences in the areas of enjoyment of human rights and the reproduction of social inequality. From these traits, it is understood that discrimination acquires a structural character in a socially complex process, where the accumulation of disadvantages is systematic and persistent over a long period of time, with "generalized effects and expressed in significant social gaps, as well as producing effects on social life, the quality of democracy, and the country's development expectations as a whole" (Solís 2017, p. 38).

On the other hand, from the jurisprudential evolution of the Inter-American Court of Human Rights, Pelletier (2014) points out that structural discrimination presents the following attributes: (a) the existence of a same group affected with common characteristics; (b) that the group is in an unreasonable disadvantage, marginalized, excluded, or vulnerable; (c) that discrimination is contextualized from historical, social, economic, and cultural perspectives; (d) that patterns of discrimination are systematic, massive, or collective in a specific geographical area, state, or region; and e) that the policy, measure, or norm is discriminatory or generates situations of unreasonable disadvantage to the group. From here, it is highlighted that structural discrimination is the result of combined factors, both the characteristics of the social group, and the various contextual dimensions that position said group in systematic social inequality.

To present a third approach, the proposal of Rodríguez (2023) is revisited, who indicates that "the appearance of naturalness or logic of discriminatory acts indeed comes from this structural dimension" (p. 300). Therefore, the author lists ten characteristics that outline the structural element in discrimination (Table 1.1).

From the extensive and detailed list of characteristics developed by Rodríguez (2023), along with the proposals of Solís (2017) and Pelletier (2014), it can be understood that discrimination is structural, stemming from the interconnectivity among cultural, social, and political elements that comprise a complex framework. This framework legitimizes, reproduces, and perpetuates discriminatory practices aimed at certain social groups over time. Therefore, it is understood that the solution to discrimination lies in the same terms in which the phenomenon presents itself, namely, structural.

If discrimination were to manifest at an individual level through concrete actions and behaviors, it would suffice to think that "if everyone became aware of the injustice underlying discriminatory practices and acted accordingly, then we would succeed in eradicating these practices" (Solís 2017, p. 33). However, as previously discussed, it is a phenomenon that transcends individual rational acts, so solutions cannot rely solely on formal and informal education, but rather through paradigm shifts, socio-symbolic reconfigurations, and, above all, proposing a structurally-based treatment

Table 1.1 Characteristics of structural discrimination

1. It is a phenomenon that operates within, but also beyond, the will and intention of individuals
2. It is a simultaneous form of social differentiation and domination
3. Discrimination is a syntax of variations that allows for explaining homologies between discriminated groups that seem disconnected
3. It is a violation of the human right to equal treatment, where the structural criterion bridges the differences between each discriminated group and affirms the consistency and cross-sectional extent of the phenomenon
4. It is a phenomenon of socio-symbolic and linguistic order within society that provides reference points of meaning in everyday life
5. It is a phenomenon of high social density and heavy materiality, where the social relations determining central aspects of collective interaction are material
6. It is a fundamental institution of the basic structure of society. It is an institution because it operates as a set of standardized norms, rules, and prohibitions that guide individuals' behavior
7. It is a political relationship, of domination, between structurally correlated groups, marked by relations of differentiation and subordination
8. It is a social pattern that tends not to let individuals escape the spaces in which they socialize and act
9. It is a phenomenon that can be overcome, if and only if the democratic state is capable of articulating a structural strategy based on a discourse of human or fundamental rights

Source Own elaboration based on Rodríguez (2023, pp. 298–300)

grounded in full access to human rights. However, there is a recognized need for a more rigorous and specific approach to discrimination. Here, intersectionality is chosen, as this approach allows for the identification of different population groups that are subject to various forms of discrimination, whether due to their age, ethnic origin, socio-cultural background, or legal status within a society.

1.5 Intersectional Discrimination

Talking about discrimination today involves conducting an analysis from an intersectional perspective, as this approach allows for the identification of different population groups subjected to various forms of discrimination, whether due to their age, ethnic origin, socio-cultural background, or legal status in society.

The starting point for the concept of intersectionality is with the American academic Kimberlé Crenshaw in the late 1980s, who coined this term to analyze the different ways in which race and gender interact in discrimination, as the phenomenon cannot be understood from a single factor or the sum of them, but in the simultaneous interaction of all dimensions (Salomé 2017). Crenshaw's proposal is a watershed moment for distinguishing the complexity of discrimination, prompting others, such as Angela Davis, to join the debate on intersectionality and enrich it with aspects of identity, multiple and changing diversities, as well as otherness (Rodríguez 2019).

Thus, the general sense of the concept of intersectional discrimination refers to "two or more sources of discrimination that, when combined, result in a qualitatively distinct inequality situation different from the sum of the parts or the forms of discrimination considered separately" (Añón 2013 in Salomé 2017, p. 271). This definition lacks highlighting the simultaneous nature, as a person may be deprived of their human rights due to the concurrence of two or more categories.

Due to the distinction of more than one factor in the prevalence of discrimination, in addition to intersectional, it has also been termed multiple, which "is found in a series of situations in which the coexistence of several motives of discrimination operating together left one in a situation of invisibility" (De Lama 2013, p. 272). This similarity in referring to a specific phenomenon of discrimination has sparked debates to point out the differences. Makkonen (cited in Rodríguez 2019) leans towards indicating that there are differences between both concepts: in multiple discrimination, different motives occur at different temporal moments, while in intersectional discrimination, the various motives interact simultaneously. In contrast, Rey Martínez (cited in Salomé 2017) indicates that both categories refer to the same phenomenon, as two or more factors of discrimination intervene simultaneously, however, intersectionality is more frequently used in the Anglo-Saxon context and multiple discrimination is commonly used in European institutions.

In the face of an ongoing debate, here intersectional discrimination is addressed based on Crenshaw's foundations, who maintains within the theoretical framework that "the intersectional domain is the result of distinct and converging processes of discrimination that, when interacting with other mechanisms of domination, such as social class status, give rise to a new dimension of disadvantage" (Rodríguez 2023, p. 200). Thus, talking about discrimination today involves analyzing from an intersectional perspective, as this approach allows for the identification of different population groups subjected to various forms of discrimination, whether due to their age, ethnic origin, socio-cultural background, or legal status in society.

Guzmán and Jiménez (2015) argue that intersectionality is an analytical tool that allows for the integration of different axes of analysis in the face of complex realities, and they present a framework to show how endogenous and exogenous variables influence how a person can be vulnerable (Fig. 1.1).

While the authors' proposal specifically addresses the analysis of violence against women, it proves to be extremely useful for this book as well. It reveals up to six levels of variables to analyze, contributing to a person experiencing discrimination and, consequently, the violation of their rights. Therefore, this approach is indispensable for shedding light on the real and deep-seated problems that arise in society. Below are the discrimination factors highlighted in the cases analyzed in this book.

1.5 Intersectional Discrimination

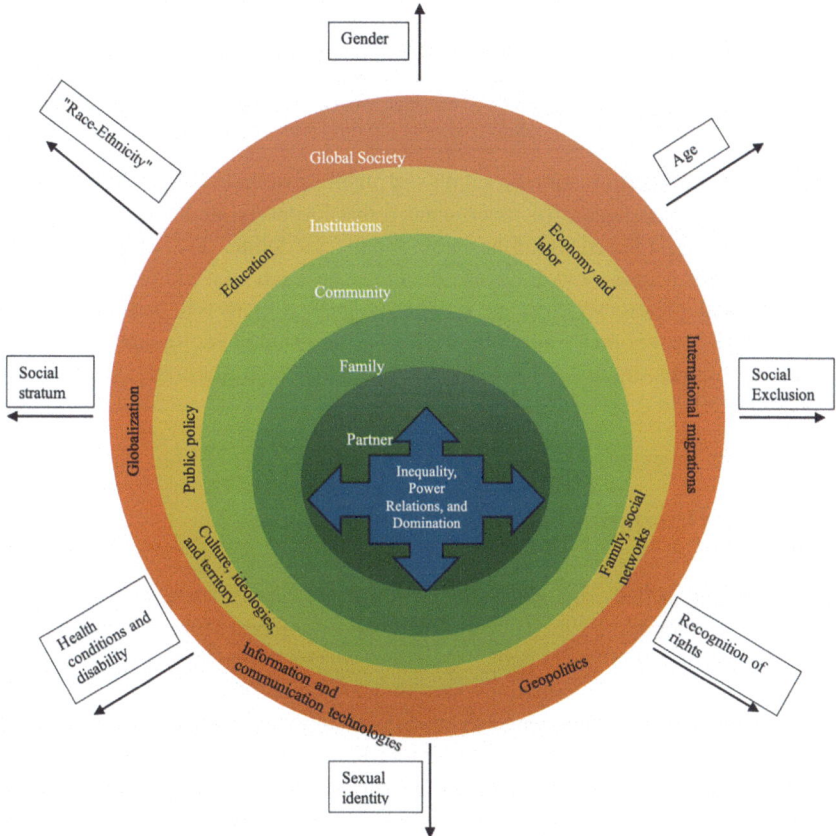

Fig. 1.1 Intersectional analysis framework for discrimination. *Source* Own elaboration based on the proposal by Guzmán and Jiménez (2015, p. 605)

1.5.1 Discrimination Based on Gender (Sex)

Gender discrimination is one of the primary factors leading to the violation of fundamental rights. According to Nuvaez (2019), it refers to practices or norms that prevent equality of interests and rights between men and women, resulting in unequal treatment. Gender pay gaps, unequal opportunities, and differential treatment are evidence of gender discrimination. Strategies such as gender parity for positions of power and decision-making have been pursued, but they have been insufficient against a structure that benefits men over women.

Specifically, this book focuses on the discrimination faced by women due to stereotypes that degrade them and undervalue their abilities, excluding them from enjoying full rights. In this regard, reference is made to the definition established by the Convention on the Elimination of All Forms of Discrimination against Women (CEDAW), which defines gender discrimination as:

> Any distinction, exclusion, or restriction based on sex that has the purpose or effect of impairing or nullifying the recognition, enjoyment, or exercise by women, irrespective of their marital status, based on equality between men and women, of human rights and fundamental freedoms in the political, economic, social, cultural, or any other sphere. (ONU Mujeres 2016, p. 1)

The effects of discrimination based on sex manifest in violence, poverty, lack of legal protection, deprivation of inheritance rights, property, and access to credit. Therefore, as established by this international guideline, the elimination of discrimination is necessary to eradicate inequality between men and women in both the public and private spheres, as well as to ensure their full participation in political, social, economic, and cultural life, by the principles of equal rights and respect for human dignity (The Office of the High Commissioner for Human Rights [OHCHR] n.d.a).

1.5.2 Age Discrimination

Age is a human attribute that has been considered a factor of social differentiation leading to discrimination, which occurs when opportunities are limited or prevented based on age (Nuvaez 2019). Whether someone is deemed too old or too young, they are placed in a vulnerable situation where they are denied the possibility to freely access or exercise their fundamental rights. Specifically, for young people, a concept has emerged that encompasses this condition, known as youngism (Francioli and North 2021). This concept refers to discriminatory elements faced by this population sector, which relate to their age as well as their appearance and identity.

On the other hand, in 1969, R. Butler first used the term "ageism," thus articulating what has always existed in various forms. It became a new concept that has contributed to making visible the different ways in which people are discriminated against based on their age. It refers to stereotypes at different stages of life laden with prejudices (Organización Panamericana de la Salud [OPS] 2021).

1.5.3 Discrimination Based on National Origin

From the conceptual construction of a Nation, it is understood that "national" refers to any person who by birth or naturalization belongs to a nation (Arredondo 2010, p. 7). In other words, national origin refers to the country or land where a person was born or where nationality was granted to them (SOS Racisme and Institut de Drets Humans de Catalunya 2019, p. 2). Seen in this light, the fact of "being part of," "belonging to," or "being linked" with a State implies the recognition, enjoyment, and exercise of a set of human rights, duties, and obligations.

Discrimination based on national origin becomes evident when nationality is positioned as a requirement for the exercise of certain rights (SOS Racisme and

Institut de Drets Humans de Catalunya 2019, p. 3). This is when the category of "foreigner," i.e., the one assigned to individuals who do not possess the nationality of the State they are in, results in unequal treatment of migrants and others in the context of human mobility. Therefore, it is important to note that the distinction between nationals and foreigners should not be considered "normal" when discussing human rights, particularly considering that every migration policy is conditioned by certain obligations of international law, including the protection of human rights (García 2012, p. 85).

1.5.4 Discrimination Based on Migratory Status

From the previous definition, it is important to distinguish between national origin and migratory status, especially considering the multiple discriminations that arise from each category. As mentioned earlier, the first term refers to the fact of being linked to a state, while migratory status corresponds to the condition of regularity or irregularity.

> This means that a person will have a regular migratory status if they have complied with the immigration provisions to enter and stay in the country; conversely, if a person has failed to comply with or ceased to comply with such provisions, their migratory status will be irregular. (Cámara de Diputados del H. Congreso de la Unión, 2024, p. 6)

From this perspective, it should be understood that the irregularity of migratory status should not be positioned as a restriction for the protection and exercise of human rights. In the case of Mexico, the Migration Law of 2011 states that among the principles on which migration policy must be based is the unrestricted respect for the human rights of migrants and foreigners regardless of their nationality and migratory status (Agencia de la ONU para los Refugiados [ACNUR] n.d.).

1.5.5 Discrimination Based on Ethnic Origin

The cultures of indigenous peoples represent a valuable cultural heritage in the world; however, these peoples constitute the poorest sector in the countries where they live and often are victims of human rights violations, including discrimination and social exclusion (International Labour Organization [ILO] 2007). Discrimination based on belonging to an indigenous people is not only based on cultural criteria such as language, customs, or dress but also on racialized physical features, such as skin tone. The latter is supported by racism as a cultural construct that legitimizes an asymmetrical relationship based on the belief that hierarchies exist based on distinctions of physical appearance (Solís et al. 2019).

The International Convention on the Elimination of All Forms of Racial Discrimination adopts a broad criterion and includes ethnic origin within racial discrimination, which it defines as:

> Any distinction, exclusion, restriction, or preference based on race, color, descent, or national or ethnic origin which has the purpose or effect of nullifying or impairing the recognition, enjoyment, or exercise, on an equal footing, of human rights and fundamental freedoms in the political, economic, social, cultural, or any other field of public life (The Office of the High Commissioner for Human Rights) [OHCHR] n.d.b).

On its part, the International Labour Organization (ILO) Convention 169 states in its third article that "indigenous and tribal peoples shall enjoy fully the human rights and fundamental freedoms without hindrance or discrimination" and urges States to take measures to safeguard the persons, institutions, property, labor, cultures, and environment of these peoples (ILO 1989). This instrument establishes a framework for the protection of indigenous peoples by international law, recognizing the value of their existence and the right to continue to exist, maintain their institutions and forms of social organization, as well as to determine the course of their development, which includes eradicating discriminatory practices based on their ethnic membership.

1.5.6 Labor Discrimination

In Chap. 4 of the Guide on International Labour Standards (2014), the ILO establishes that any state that ratifies Convention 111[3] is obligated to carry out national policies that promote equal opportunity and treatment, to eliminate any type of discrimination in access to professional training, employment admission, and various occupations, as well as in working conditions.

The document expresses six recommendations aimed at equal opportunities and treatment in hiring, training, promotions, job retention, and working conditions, as well as in areas of safety and hygiene, collective bargaining, affiliation with employers' or workers' organizations, participation in trade union matters, both in public and private spheres (OIT 2014). Likewise, three exceptions are explicitly mentioned: "based on the qualifications required for a job; that can be justified by the protection of the security of the State; having the character of protection or assistance measures" (OIT 2014, pp. 35–36). In this way, the ILO promotes the exercise and protection of labor rights and, at the same time, indicates exceptions that prohibit equal opportunities and treatment in favor of vulnerable or disadvantaged population groups. For example, having a hiring policy for older persons to help a sector that can hardly find employment would also be absurd to prohibit using age criteria at all times, as this would prevent protecting minors from work (Soberanes 2022, p. 276).

---

[3] In Chap. 4, reference is made to Convention No. 111 concerning Discrimination (Employment and Occupation), of 1958.

As argued, discrimination is sustained and reproduced through power relations, evident in the domination-oppression dyad, and the problem that arises for the study of this relationship corresponds to the methodological realm, as the variables that need to be observed can be many and even dissimilar. Faced with this theoretical-methodological reality, intersectionality emerges as a tool that offers the possibility of analyzing discrimination from quantitative and qualitative perspectives, thus contributing to a better analysis of social inequalities and the violation of people's rights.

1.6 Conclusion

As noted in the development of this chapter, the term "discrimination" has various meanings, but the one of interest for this book is where differential treatment and actions taken to the detriment of a person or group of people lead to violating human rights. From this perspective, discrimination should be understood as the denial of fundamental rights based on unjustifiable grounds such as age, gender, religion, place of origin, ideologies, sexual preferences, and others.

In this sense, discrimination is rooted in relationships of domination and oppression, where certain groups exert power over others, perpetuating inequality and social exclusion. These relationships are sustained over time, supported by social and cultural mechanisms that reinforce the hegemony of some over others, such as aporophobia, stigma, and prejudice. It is essential to focus attention on these mechanisms to make visible and analyze the interaction processes between individuals, and thereby the systems of intersubjective relationships that generate asymmetries of domination and exclusion.

On the other hand, it is important to emphasize the human rights approach to combating discrimination. As mentioned in this chapter, it is not only unequal treatment that leads to discrimination but also the denial of fundamental rights. Therefore, it is essential to adopt this approach to ensure the free exercise of all individuals or population groups' rights, especially those subjugated by social and cultural mechanisms that have placed them in vulnerable conditions.

Two forms of discrimination were also highlighted: structural and intersectional. The first emphasizes systemic processes that affect certain groups over a long period, leading to discrimination becoming normalized and accepted. In the case of intersectional discrimination, it is recognized that a person may be discriminated against based on multiple simultaneous factors, such as gender, age, ethnicity, and place of origin, among others. Both types of discrimination present complexities in terms of detection, analysis, and intervention, but recognizing them is essential to implementing actions that ensure human rights.

Finally, as presented in this chapter, it is acknowledged that education and public policies are essential for eradicating discrimination. Education is vital for fostering equality and respect for human rights for all individuals and population groups. Meanwhile, public policies must promote and ensure inclusion, both in written laws and in the mechanisms that enforce such policies. From these and other factors, structural changes can be considered to eliminate discrimination and contribute to the construction of peaceful and inclusive societies.

References

Agencia de la ONU para los Refugiados. (n.d.). *Buena práctica 13: no discriminación por situación migratoria.* https://acnur.org/fileadmin/Documentos/Proteccion/Buenas_Practicas/9219.pdf

Arredondo, F. (2010). *Personas físicas nacionales y extranjeras.* Régimen jurídico (2ª. ed.). Colección Colegio de Notarios del Distrito Federal.

Banco Mundial. (2018, 19 de septiembre). *Según el Banco Mundial, la pobreza extrema a nivel mundial continúa disminuyendo, aunque a un ritmo más lento* [Comunicado de Prensa N.º 2019/030/DEC-GPV]. https://www.bancomundial.org/es/news/press-release/2018/09/19/decline-of-global-extreme-poverty-continues-but-has-slowed-world-bank

Cámara de Diputados del H. Congreso de la Unión (2024). Ley de Migración. Nueva Ley publicada en el Diario Oficial de la Federación el 25 de mayo de 2011. https://www.diputados.gob.mx/LeyesBiblio/pdf/LMigra.pdf

Comisión Económica para América Latina y el Caribe. (2013). *Consenso de Montevideo sobre población y desarrollo.* Comisión Económica para América Latina y el Caribe. https://www.cepal.org/es/publicaciones/21835-consenso-montevideo-poblacion-desarrollo

Comisión Económica para América Latina y el Caribe. (2015). *Guía Operacional para la Implementación y el Seguimiento del Consenso de Montevideo Sobre Población y Desarrollo.* CEPAL. https://www.cepal.org/es/publicaciones/38935-guia-operacional-la-implementacion-seguimiento-consenso-montevideo-poblacion

Comisión Nacional de los Derechos Humanos. (2012). *La discriminación y el derecho a la no discriminación.* CNDH. https://www.cndh.org.mx/sites/all/doc/cartillas/2015-2016/43-discriminacion-dh.pdf

Cortina, A. (2017). *Aporofobia, el rechazo al pobre. Un desafío para la democracia.* Paidós.

De Lama, A. (2013). Discriminación múltiple. *Anuario de derecho civil*, 66(1), 271–320. https://revistas.mjusticia.gob.es/index.php/ADC/article/view/3715

Francioli, S. & North, M. (2021). Youngism: The content, causes, and consequences of prejudices toward younger adults. *Journal of Experimental Psychology General, 150*(12), 1–22. https://doi.org/10.1037/xge0001064

García, T. (2012). El estatus de extranjería en México. *Propuesta de reforma migratoria,* 45(133), 55–91. https://revistas.juridicas.unam.mx/index.php/derecho-comparado/article/view/4734/6085

Giacaglia, M. (2002). Hegemonía. Concepto clave para pensar la política. *Tópicos. Revista de Filosofía de Santa Fe*, (10), 151–159. https://doi.org/10.14409/topicos.v0i10.7430

Godelier, M. (1986). *La producción de los grandes hombres. Poder y dominación masculina entre los Baruya de Nueva Guinea.* Akal.

Goffman, E. (2006). *Estigma. La identidad deteriorada.* Amorrortu.

Gramsci, A. (1975). *Cuadernos de la cárcel* (tomo 1). Ediciones Era.

Guzmán, R. y Jiménez, M.L. (2015). La Interseccionalidad como instrumento analítico de interpelación en la violencia de género. *Oñati Socio-legal Series,* 5(2), 596–612. http://ssrn.com/abstract=2611644

International Labour Organization. (1989). *Indigenous and Tribal Peoples Convention, 1989 (No. 169).* ILO. https://normlex.ilo.org/dyn/nrmlx_en/f?p=NORMLEXPUB:55:0::NO::P55_TYPE%2CP55_LANG%2CP55_DOCUMENT%2CP55_NODE:REV%2Cen%2CC169%2C%2FDocument

References

International Labour Organization. (2007). *Newsletter 2007. The ILO and the indigenous and tribal people. Theme: Discrimination*. ILO. https://www.ilo.org/wcmsp5/groups/public/@ed_norm/@normes/documents/publication/wcms_100542.pdf

Naciones Unidas. (2001). *Conferencia Mundial contra el Racismo, la Discriminación Racial, la Xenofobia y las Formas Conexas de Intolerancia*. NU. https://www.un.org/es/conferences/racism/durban2001

Naciones Unidas. (2018). *La Agenda 2030 y los Objetivos de Desarrollo Sostenible: una oportunidad para América Latina y el Caribe (LC/G.2681-P/Rev.3)*. NU.

Naciones Unidas. (n.d.). *La Declaración Universal de los Derechos Humanos*. NU. https://www.un.org/es/about-us/universal-declaration-of-human-rights

Nuvaez, J. (2019). La discriminación laboral en razón del género y la edad en Colombia. *Revista Arbitrada Interdisciplinaria Koinonía*, 4(7), 308–320. https://doi.org/10.35381/r.k.v4i7.207

ONU Mujeres. (2016). *La CEDAW, Convención sobre los Derechos de las Mujeres*. ONU Mujeres. https://mexico.unwomen.org/es/digiteca/publicaciones/2016/01/la-cedaw-convecion-derechos-de-las-mujeres

Organización Internacional del Trabajo. (2014). *Guía sobre las normas internacionales del trabajo*. OIT. https://www.ilo.org/wcmsp5/groups/public/---ed_norm/---normes/documents/publication/wcms_246945.pdf

Organización Panamericana de la Salud. (2021). *Informe mundial sobre el edadismo*. OPS. https://iris.paho.org/handle/10665.2/55871.

Pelletier, P. (2014). La "discriminación estructural" en la evolución jurisprudencial de la Corte Interamericana de Derechos Humanos. *Revista IIDH*, 60, 205-215. https://www.corteidh.or.cr/tablas/r34025.pdf

Real Academia Española. (2023). *Diccionario de la lengua española*. Recuperado el 20 de noviembre del 2023, de https://www.rae.es/

Rodríguez, J. (2023). *Una teoría de la discriminación*. Universidad Autónoma Metropolitana-Iztapalapa.

Rodríguez, V. (2019). La discriminación interseccional en el discurso Jurídico. *Nuevo Derecho*, 15(25), 70–87. https://revistas.iue.edu.co/index.php/nuevoderecho/article/view/1235/pdf

Salomé, L. (2017). La discriminación y algunos de sus calificativos: directa, indirecta, por indiferenciación, interseccional (o múltiple) y estructural. *Pensamiento constitucional*, 22(22), https://revistas.pucp.edu.pe/index.php/pensamientoconstitucional/article/view/19948/19969

Soberanes, J. (2022). La discriminación en las convocatorias laborales. *Revista latinoamericana de derecho social*, 35, 271–296. https://doi.org/10.22201/iij.24487899e.2022.35.17279

Solís, P. (2017). *Discriminación estructural y desigualdad social. Con casos ilustrativos para jóvenes indígenas, mujeres y personas con discapacidad*. SEGOB, CONAPRED y CEPAL.

Solís, P., Krozer, A., Arroyo, C. y Güemez, B. (2019). *Discriminación étnico-racial en México: una taxonomía de las prácticas. Documento de Trabajo # 1. Proyecto sobre Discriminación Étnico Racial en México (PRODER)*. El Colegio de México https://discriminacion.colmex.mx/wp-content/uploads/2019/08/dt1.pdf

SOS Racisme y Institut de Drets Humans de Catalunya. (2019). *Discriminación racial, discriminación por origen nacional: el caso de las leyes de migración y/o extranjería*. IDHC. https://www.idhc.org/es/publicaciones/discriminacion-racial-discriminacion-por-origen-nacional-el-caso-de-las-leyes-de-migracion-y-o-extranjeria.php

The Office of the High Commissioner for Human Rights. (n.d.a). *CEDAW in your daily life*. OHCHR. https://www.ohchr.org/es/treaty-bodies/cedaw/cedaw-your-daily-life

The Office of the High Commissioner for Human Rights (n.d.b). *International Convention on the Elimination of All Forms of Racial Discrimination*. OHCHR. https://www.ohchr.org/en/instruments-mechanisms/instruments/international-convention-elimination-all-forms-racial

Vargas, Y., Santana, C., Torres, E., y Gutiérrez, S. (2020). Comparación de marcos conceptuales de la teoría de la discriminación de J. Rodríguez Zepeda y Adela Cortina. *Sincronía*, (77), 450–462.

Chapter 2
Current Context and Challenges Regarding Discrimination in Mexico

In recent decades, discrimination has been increasingly recognized as a problem of social inequality that entails the denial of rights, while the prevalence of discriminatory practices towards different vulnerable groups in Mexico has been acknowledged. Building upon the legal and institutional framework in the fight against discrimination, the objective of this chapter is to provide a current overview of the experiences and discriminatory practices faced by the international migrant population, indigenous people, youth, and older persons in Mexico.

To achieve this, we first outline the incorporation of discrimination into the public agenda and its acknowledgment as a structural issue that demands legal mechanisms and public action. Subsequently, a descriptive analysis is presented based on the processing of data from the National Survey of Discrimination (ENADIS) 2017 and 2022, which allows us to highlight the persistence of prejudices translating into multiple discriminations and denial of rights in various spheres.

2.1 Discrimination in the Public Agenda in Mexico

The State's involvement in combating discrimination in Mexico has a recent history. Until the year 2000, the official position of the Mexican government was that there was no discrimination, under the belief that it was only based on race and, if the ideology of mestizaje, based on the rational and cultural blending that provided identity to the Mexican population, was considered valid, then there was no discrimination (Rincón 2005, pp. 7–8).

Previously, numerous social movements, political groups, and experts from various backgrounds had already challenged the myth of a unitary nation, making evident the exclusion, segregation, and limitation of rights for different social groups due to stigmas and prejudices. At the same time, they had advocated for a state policy aimed at combating discrimination.

Discrimination was included on the public agenda until the year 2000, within the context of political alternation at the national executive level. The legal and institutional fight against discrimination had a clear starting point with the establishment of the Citizens' Commission for Studies against Discrimination, installed in February 2001 and chaired by Gilberto Rincón Gallardo (Rodríguez 2018). It was composed of representatives from major political parties, legislators, public officials, representatives of organizations, academics, and specialists in the field (Rincón 2005).

The Commission had the opportunity "to articulate a comprehensive program to combat discrimination while simultaneously issuing a powerful call to various relevant social and political actors that would be crucial for legislative processes and institutional development in antidiscrimination matters" (Rodríguez 2018, p. 51). It emphasized the need to formulate the public agenda for non-discrimination by the international framework of human rights and to avoid assistance-oriented approaches.

Among the actions of the Commission, highlights include the first diagnosis of the discrimination situation at the national level and the promotion of the Federal Law initiative to Prevent and Eliminate Discrimination (LFPED), as well as the proposal for the creation of a National Council to Prevent Discrimination focused on coordinating the efforts of the Mexican state in the fight against discrimination (Raphael 2012, p. 44).

In addition to the remarkable work of the Commission and the voices of various sectors against discrimination, the orientation of the agenda towards equal treatment also echoes international treaties. Mexico has ratified more than fifty international instruments where it commits, in various ways, to combat discrimination (Raphael 2012).

Regarding the legal framework, the legal fight against discrimination in Mexico involves specific protections for individuals or groups of people who are vulnerable to acts of disdain, exclusion, or even violence (Rincón 2005). In this framework, legislative changes are significant as they consider the right to non-discrimination as universal and applicable to all individuals. Likewise, and without denying the above, it demands specific policies and treatments for different groups vulnerable to discrimination, under the premise that discrimination is a form of inequality that occurs through intergroup relations, between discriminating and discriminated groups and individuals (Rodríguez 2018).

Due to its significance, constitutional reforms and the creation of a General Law in the field stand out. In the first instance, the constitutional reform in 2001 to Article 1, which enshrines the mandate of non-discrimination, identifies various grounds for discrimination and recognizes that these are acts that undermine the dignity and rights of individuals.

Another constitutional change in 2011, regarding the denomination of the first chapter, first title, was renamed "Of Human Rights and Their Guarantees", which elevates the concept of Human Rights to constitutional rank and leaves behind the idea of Individual Guarantees. This change establishes the obligation of the State to promote and guarantee Human Rights, including non-discrimination (Raphael

2012). From these important reforms, the right to non-discrimination is enshrined in the Constitution as follows:

> All discrimination based on ethnic or national origin, gender, age, disabilities, social status, health conditions, religion, opinions, sexual preferences, marital status, or any other that undermines human dignity and aims to nullify or impair the rights and freedoms of individuals is prohibited (Secretaría de Gobernación 2011, p.1).

The Federal Law to Prevent and Eliminate Discrimination was published on June 11, 2003, in the Official Gazette of the Federation (Secretaría de Gobernación 2003). It is a legal instrument that stems from incorporating discrimination into the public agenda by recognizing it as a social issue requiring attention from the Mexican state through concrete actions. This law prohibits discriminatory practices aimed at nullifying the exercise of rights while identifying discriminatory acts and guiding state action on the matter.

Since then, Mexico has had a top-level legal instrument focused on preventing and eliminating all forms of discrimination in terms of Article 1 of the Constitution, while also establishing guidelines for public bodies and federal authorities to carry out positive and compensatory measures aimed at achieving equal opportunities. The law particularly protects vulnerable groups while compensating and empowering certain groups to access opportunities that have been restricted due to discriminatory practices. This means simultaneously providing protection and taking action (Rincón 2005).

Mexican legislation demands a comprehensive approach to discriminatory practices by proposing the elimination of all barriers hindering the exercise of rights and access to equal opportunities for all individuals. In this regard, individuals belonging to historically stigmatized and excluded groups have legal protection against discriminatory practices, without negating the universal nature of the norm (Rincón 2005).

The goal is not to generate laws with exclusive applications and perpetuate differences, but rather to build egalitarian relationships and ensure that the constitutional norm serves the social groups most vulnerable to discriminatory practices. This is intended to provide all individuals with the opportunity to access social conditions conducive to their social inclusion and holistic personal development (Rodríguez 2004).

Institutionally, the creation of CONAPRED stands out, which is a state body created by the Federal Law to Prevent and Eliminate Discrimination. CONAPRED began operations in 2004, with the institutional and social commitment to promote a culture of justice and equity, respect the right to non-discrimination, and develop actions to prevent and eliminate discrimination (Cámara de Diputados 2003). It intends to contribute to the construction of a collective order "that recognizes differences and coexistence in diversity as elements that contribute to the construction of a high-quality democratic order" (Rincón 2004, p. 5).

Through the Department of Guidance, CONAPRED provides legal guidance to all individuals requesting the intervention of this Council regarding alleged discriminatory acts, omissions, or social practices attributed to individuals, entities (natural or legal persons), federal public servants, and federal public authorities. This department advises petitioners on their rights and means of defense, and in cases not within the Council's jurisdiction, it directs them to the appropriate authority (CONAPRED 2019, p. 27).

The process of addressing alleged acts of discrimination involves various procedures depending on each case. To ensure greater effectiveness in addressing the issue, various activities may be undertaken with authorities and individuals to resolve the problem before filing a complaint. When possible and depending on the case, mediation, sending referral letters to state agencies, communications and letters of prevention and exhortation, as well as requests for collaboration are pursued (CONAPRED 2020). This is done to prevent petitioners from later filing a complaint and initiating a process that would require more time for attention, resolution, and follow-up.

Over the years, the Council has positioned itself as a federal-level service, as requests for guidance have remained stable over time, with a significant increase in the last recorded year (Fig. 2.1). This indicates, on one hand, the persistence of acts undermining the dignity and access to rights of vulnerable groups, but at the same time, a more conscientious society regarding the role of CONAPRED as a service provider to assert their rights.

Regarding complaints, records have been decreasing, as a result of the previously mentioned attention processes, as well as the existence of entities at the state and local levels to prevent and eliminate discrimination, which handles cases

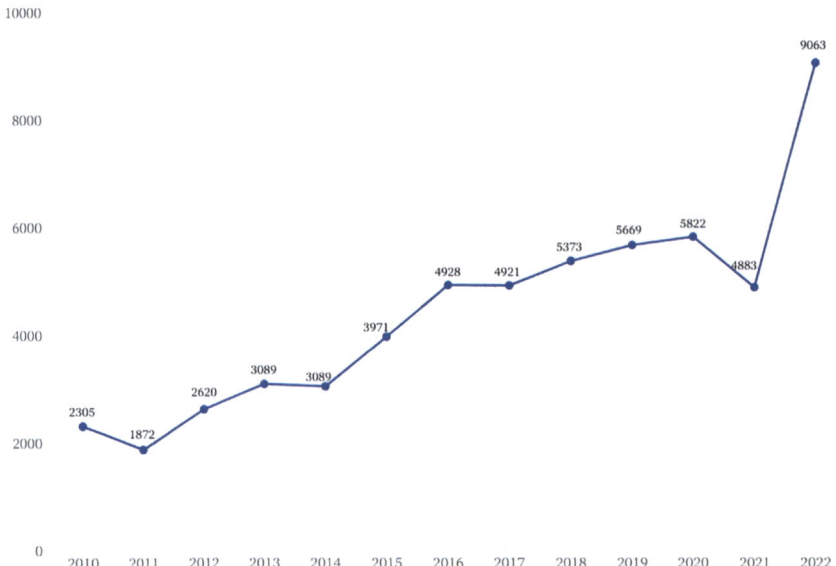

Fig. 2.1 Consultations and guidance provided by CONAPRED, 2010–2022. *Source* Own elaboration based on the annual reports of CONAPRED (n.d.a, 2022, 2023)

at the state and local levels. However, it is worth noting the prevalence of discriminatory practices recorded by the Council in the last five years, as the causes of discrimination in complaints remain consistent over time regarding disability, health condition, sexual orientation, gender, physical appearance, and age (Table 2.1). These are records resulting from formal and institutional processes at the federal level, but they demonstrate the persistence of inequalities, as well as the interest in asserting the right to non-discrimination.

The fight against discrimination not only involves promoting the protection of specific groups. It requires the state to promote and encourage compensations and measures for the social integration of certain discriminated groups, which demands the presence of affirmative actions to achieve social cohesion and respect for differences (Rincón 2005, p. 8).

In that vein, CONAPRED participates in the design and implementation of the National Program for Equality and Non-Discrimination (PRONAIND). The first program was published in 2014 and has since been the instrument guiding the course of action to combat institutionalized discriminatory practices based on the transversality of the right to equality and non-discrimination (CONAPRED n.d.b). The PRONAIND 2021–2024 sets forth a series of priority objectives aimed at eradicating discriminatory practices in the following areas: health, education, employment, social security, security, and justice (Secretaría de Gobernación 2021, p. 29). To this end, strategies and actions are established, the scope of which to dismantle institutionalized discriminatory practices and their impact on reducing inequality gaps is yet to be seen.

Table 2.1 Main causes of discrimination in complaints handled by CONAPRED, 2017–2022

2017	2018	2019	2020	2021	2022
• Disability	• Disability	• Health condition	• Health condition	• Disability	• Disability
• Health condition	• Health condition	• Disability	• Disability	• Health condition	• Health condition
• Physical appearance	• Gender	• Sexual orientation	• Gender	• Gender	• Gender
• Pregnancy	• Sexual preference or orientation	• Gender	• Sexual orientation	• Sexual orientation	• Sexual orientation
• Sexual preference or orientation	• Pregnancy	• Pregnancy	• Age	• Age	• Physical appearance
• Gender	• Physical appearance	• Sex	• Pregnancy	• Physical appearance	• Pregnancy
• Age	• Age	• Physical appearance	• Physical appearance	• Pregnancy	• Age
• National origin	• Gender identity	• Age	• Other	• Other	• National origin
• Gender identity	• National origin	• National origin	• Legal status	• Gender identity	• Gender identity
• Social status	• Family situation	• Gender identity	• Gender identity	• Legal status	• Legal status

Source Own elaboration based on CONAPRED (2018, 2019, 2020, 2021, 2022, 2023)

Based on the brief overview presented here, it is clear that institutional steps toward building a society free from discrimination have been significant, as it is recognized that equality of treatment is built upon a democratic and rule-of-law state (Raphael 2012, p. 25). This requires joint efforts to create conditions that allow for a mode of coexistence based on the recognition of rights for all individuals regardless of their belonging to vulnerable social groups.

However, discriminatory practices are deeply rooted in Mexican society, contributing to maintaining a social order that normalizes unequal treatment and opportunities. Hence the importance of CONAPRED's role in generating educational and outreach materials to promote a culture of inclusion and equal rights, as well as efforts to document and monitor the persistence of perceptions, prejudices, and denial of rights associated with various vulnerable groups, as outlined below.

2.2 Discrimination Overview: Actors, Practices, and Denial of Rights

Understanding and contextualizing the processes of discrimination faced by vulnerable groups discussed in this book requires an approach to the dynamics that are part of this issue, specifically referring to situations that denote the denial of rights and experiences of discrimination that generate specific challenges for these populations.

To achieve this, the ENADIS emerges as a valuable resource for dimensioning this issue in Mexico. This survey, conducted by the INEGI in collaboration with the CONAPRED and the CNDH, allows for the measurement of diverse aspects, including the perception of respect for rights, processes of social inequality by condition, as well as the experiences and areas where these practices are visible and occur (INEGI 2023).

ENADIS was first conducted in 2017, focusing on indigenous people, people with disabilities, people with religious diversity, older persons, women, girls and boys, adolescents, and young people. In 2022, the latest round of data collection through this survey took place, observing methodological and conceptual adjustments that show progress in offering a more comprehensive and complex overview of discrimination in Mexico, as other population groups and dimensions of analysis are added, possibly not previously incorporated because they were not visible and/or were not part of the discussion agenda for the attention of these groups.

Among the main changes in the study populations, we identify the inclusion of other sectors of interest such as Afro-Mexicans, domestic workers, and migrants, thus promoting understanding and recognition of the current situation of the groups prioritized by the United Nations for Human Rights, as they are minorities whose rights lack greater protection (NU 2023).

2.2 Discrimination Overview: Actors, Practices, and Denial of Rights

In this sense, ENADIS becomes a tool that allows for the visibility of expressions of discrimination to implement corrective actions, as well as contributing to the design of public policies that contribute to the construction of tolerant, inclusive, peaceful, and non-discriminatory societies (INEGI 2023).

As part of its methodological design, this survey gathers socio-demographic information from members of the surveyed households, from which a person aged 18 or older is randomly selected to assess their perception and experiences of discrimination. The resource we use for our analysis is the questionnaire of the modules applied to household members, whose socio-demographic characteristics distinguish them as part of the populations focused on in this research.

Among the various variables explored by this survey, we focus on (i) perception of respect for their rights, (ii) specific problems of each group, (iii) prejudices and stereotypes, (iv) denial of rights, (v) situations of discrimination experienced in the last five years; as well as (vi) the reasons and (vii) areas where these practices have occurred in the last year. The descriptive analysis we present is conducted from a comparative perspective that, between the years 2017 and 2022, allows us to distinguish the prevalence of discrimination to reflect on and question the advances and/or setbacks observed in this matter.

For this analysis, the following specifications are considered to identify, as far as possible, the population groups that share characteristics with the specific cases we develop in the following sections:

- The international migrant population includes individuals born in another country, regardless of whether they also have Mexican nationality. In this case, the data corresponds solely to the year 2022, as it's important to recall that the ENADIS did not include this population group in 2017.
- The indigenous population includes all individuals who identify as indigenous because they speak an indigenous language, were born in an indigenous community, have parents or grandparents who speak an indigenous language and/or belong to an indigenous community, as well as individuals who are recognized by a community or who identify themselves as such based on their customs or traditions.
- The youth population corresponds to individuals aged 18–24 who, at the time of the interview, were either studying and/or had a job, even if they had not worked in the last week.[1]
- In the case of older persons, individuals aged 60 and above are considered.

So, the results of the ENADIS show particularities for each population group. Regarding the *perception of rights*, it is observed that, in both years, that is, in 2017 and 2022, the indigenous population recorded the highest percentages of those who perceived that their rights were "somewhat" respected (40.2% and

---

[1] The official Mexican guideline indicates that the age range to establish youth is between 12 and 29 years old (IMJUVE 2017). However, for the purposes of this analysis, ages from 18 to 24 are considered, which correspond to legal age in Mexico and the group susceptible to enrollment in university level.

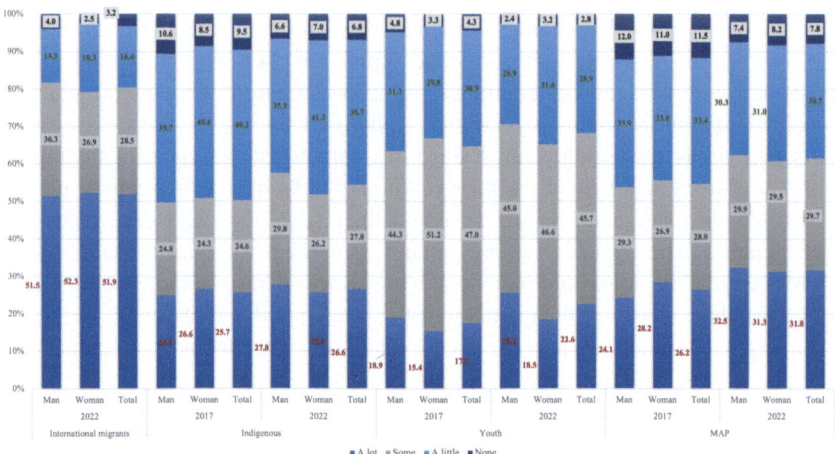

Fig. 2.2 Perception of the population belonging to vulnerable groups regarding the respect for their rights, 2017–2022. *Source* Own elaboration based on ENADIS (2017, 2022)

38.7%, respectively). In this same concept, older persons also stand out (33.4 and 30.7%) (Fig. 2.2).

Contrary to these populations, young people report a moderately favorable perception, with approximately half of them indicating that their rights are "somewhat" respected: 47.0% and 45.7% in 2017 and 2022, respectively. Finally, 51.9% of the international migrant population considers that their rights are "very much" respected. Regarding this latter population, it is worth noting that the situation of migrants in transit, asylum seekers, and even returned populations—migration modalities not captured by ENADIS—is different due to the systematic violation of their rights through processes of militarization, containment, detention, and deportation promoted by the Mexican state (Centro para la Justicia y el Derecho Internacional [CEJIL] 2019).

In 2022, differentiation by sex shows some variations. For instance, among the indigenous population, the proportion of women who perceive that their rights are "somewhat" respected is higher (41.2%), reflecting the persistence of gender gaps and the intersection of multiple inequalities that place them in particularly difficult conditions (Consejo para Prevenir y Eliminar la Discriminación de la Ciudad de México [COPRED] 2021).

While the enjoyment and full respect of human rights are closely related to the specific ***problems*** of these populations, a constant among the four groups is the difficulty and lack of opportunities to find employment, particularly considering that job creation and labor market conditions in Mexico are neither sufficient nor suitable for the demand of the population that wants and needs to work (Centro de Investigación en Política Pública 2022). Therefore, as outlined in the cases addressed in this text, the right to work is among the main rights violated for these populations, particularly when the access gap widens due to processes of discrimination.

Regarding this, the international migrant population, unlike the other three groups, recognizes the lack of employment as the main issue (27.0%), followed by discrimination based on their national origin or coming from another place (17.4%), abuse by authorities (16.1%), and lack of attention from immigration authorities (13.4%) (Fig. 2.3). This constitutes a set of situations indicating rejection towards what is considered the "other" and perceived as different while highlighting the need to promote intercultural competencies among public officials involved in the care processes for this population.

Among the indigenous population, employment also emerged as the main issue in both years (20.9 and 18.5%), a situation closely related to the lack of economic resources (16.1 and 16.9%). In this period, we observed a decrease in the percentage of the population distinguishing the lack of support schemes provided through social programs as their main difficulties (from 15.8 to 12.6%), possibly reflecting the notion of welfare adopted by the current Mexican government. Finally, the recognition of the problem of discrimination based on appearance and/or speaking an indigenous language occurs because they are stigmatized in various ways by associating this population with the notions of poverty, backwardness, and ignorance (Comisión Nacional para el Desarrollo de los Pueblos Indígenas [CDI] 2006).

Among young people, the main issues in both years are related to addictions (32.7 and 31.9%) and the lack of opportunities to continue studying (21.9 and 19.9%). The lack of employment ranks third (Fig. 2.3). Nevertheless, it is important to highlight the emergence of the so-called "youngism" as a process of age-based labor discrimination against young people (Francioli and North 2021). Addressing this scarcely visible issue represents a particular challenge, especially in the face of adultcentrism as a system that disadvantages this population group (Heatley 2021).

In the case of older persons, in both years, the main issue is that their pension is not enough to cover their basic needs (28.5 and 36.3%). It is worth noting the distinction between contributory and non-contributory pensions. In the former case, the lack of resources among this population is exacerbated considering that the vast majority of people do not have access to this type of pension (Félix et al. 2023), while among non-contributory pensions, initiatives such as the Universal Pension Program for Older persons become visible, aimed at ensuring that this population receives bi-monthly financial support to improve their living conditions (Gobierno de México 2025).

These issues also have close ties to prejudices, as they involve attitudes toward certain individuals or groups that influence discriminatory decisions and acts toward them (Pérez & Oliver 2020). For example, international migrants who agree that most people believe that individuals from other places cause various social problems (29.2%) are supported by observed attitudes and the denial of their rights, particularly in the presence of other prejudices they agree with, including the belief that they take away jobs from nationals (28.6%) and/or reject the customs of the place they arrive at (28.8%) (Fig. 2.4).

Among the indigenous population, the prejudices addressed by ENADIS relate to the issue of contempt, considering them as undervalued individuals (75.6 and 72.5%), or underestimating the development and use of their capacities and skills to

Fig. 2.3 Main issues considered by population group, 2022. *Source* Own elaboration based on ENADIS (2022)

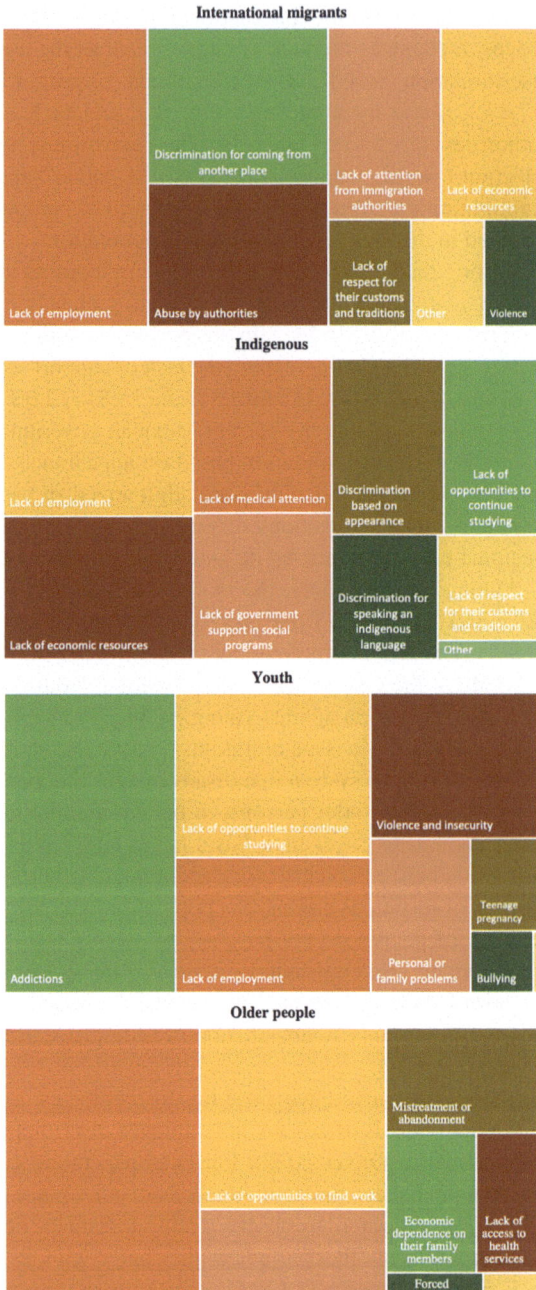

2.2 Discrimination Overview: Actors, Practices, and Denial of Rights

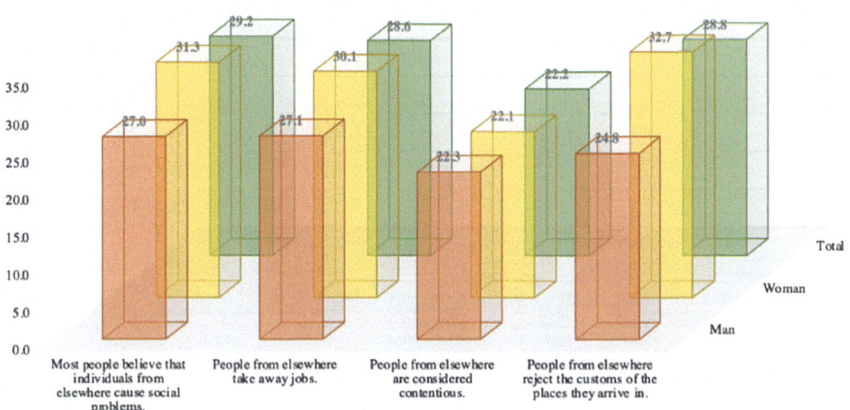

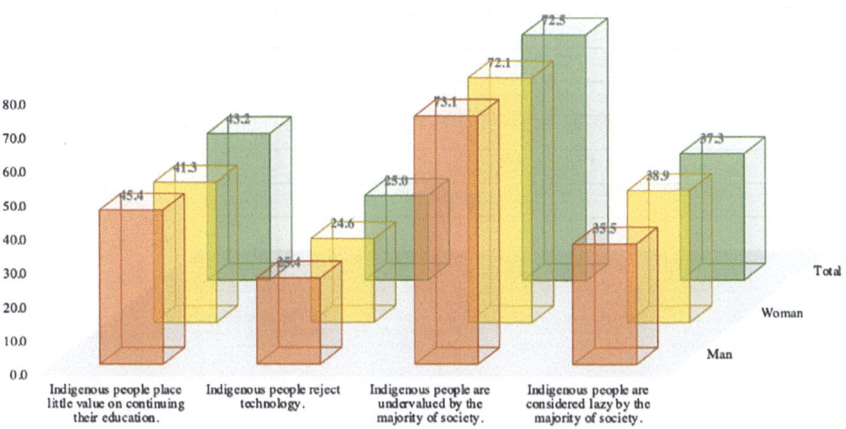

Fig. 2.4 Population agreeing with prejudices attributed to their population group, 2022. *Source* Own elaboration based on ENADIS (2022)

continue studying (43.0 and 43.2%), to work (43.7 and 37.3%), and even to use and take advantage of technology (29.0 and 25.0%), which, as we will see in the section focused on this population, responds to racism as a form of distinction combined with other expressions of exclusion and multiple discrimination (Gracia and Horbath, 2019).

For the young population, we observe that social construction results in prejudices that consider them lazy, as they neither study nor work (80.3 and 76.3%). This process of stigmatization also reproduces attitudes aimed at devaluing their opinions (75.4

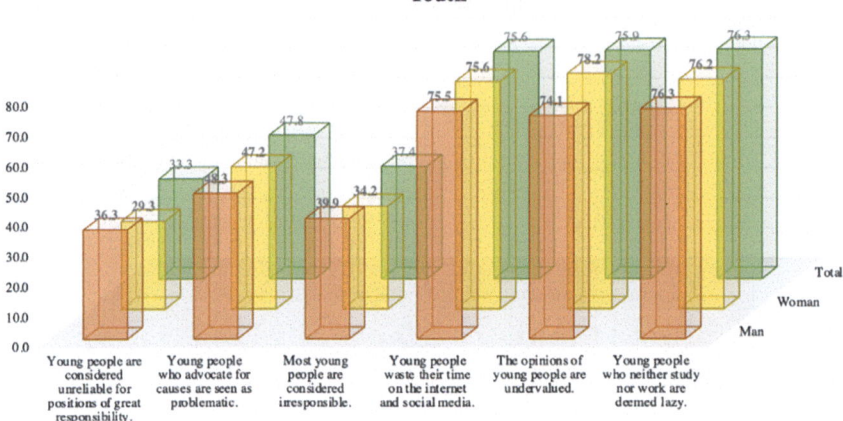

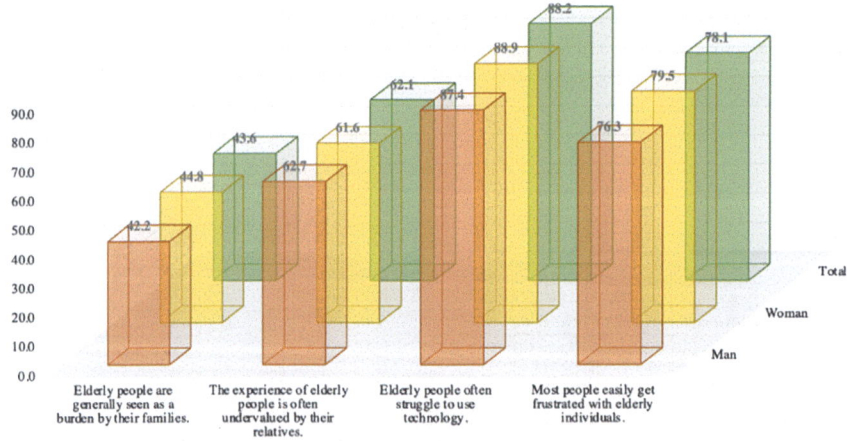

Fig. 2.4 (continued)

and 75.9%), or multiple criticisms for the time spent on and lost using social networks (60.8 and 75.6%).

Finally, the main prejudice older persons agree with is that they find it difficult to use technology (89.4 and 88.2%), which in a digitized society can translate into barriers to accessing information and denial of rights. In second place are those who agree that people easily get frustrated with them (82.0 and 78.1%), leading to a range of practices that result in mistreatment and neglect of their needs (Fig. 2.4).

Now, regarding the denial of rights in the last five years, the percentage of migrant individuals who reported the denial of at least one of their rights is lower (20.1%) compared to what was reported by the indigenous population (24.2%), young people (22.5%), and older persons (20.9%) (Fig. 2.5). The breakdown by type of rights

2.2 Discrimination Overview: Actors, Practices, and Denial of Rights

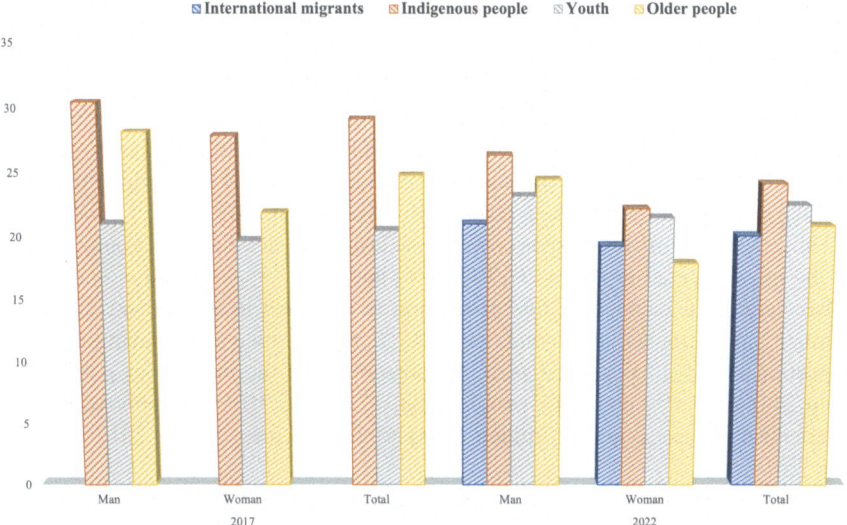

Fig. 2.5 Denial of rights by population group and gender in, the last five years 2017 and 2022. *Source* Own elaboration based on ENADIS (2017, 2022)

reveals that 38.1% of them mentioned being denied the possibility of accessing or receiving support from social programs, a proportion that is even higher among women (45.9%). Following this situation is the denial of credits, loans, or cards (34.0%), in this case, the proportion of men (42.0%) whose right to banking was violated is higher. In third place, we highlight the violation of the right to work, as 21.8% of this population specified that they were denied the opportunity to obtain a job or a promotion.

In the group of the indigenous population, there is a decrease in the number of people who reported being denied any of their rights, going from 29.2 to 24.2% from 2017 to 2022. In both years, the right to health represents one of the most violated rights (51.7 and 41.7%), followed by the right to social protection granted through access to social programs (38.2 and 47.8%) and, therefore, the right to access government office services (29.7 and 27.9%). In this case, as with the migrant population, the diversity and cultural richness of our country require the generation of intercultural competencies and skills that contribute to the construction of inclusive and tolerant societies.

Among young people, there was an increase during this period, although minimal (from 20.5 to 22.5%), indicating the persistence of discrimination against this population. During 2017, individuals who reported denial of any of their rights during the last five years specified access to social programs (35.8%), medical care (32.2%), as well as the right to work (26.6%) more often. We emphasize this last aspect because it recorded a higher proportion in 2022 (32.9%), allowing us to visualize the persistence and increasingly greater labor discrimination experienced by this population group.

For older people, there is a decrease in those who reported experiences of rights denial is distinguished (from 24.8 to 20.9%). While in 2017 the main situation responded to denial of access to social programs (40.6%), in 2022 the violation of the right to health stands out (42.5%), with a proportion that shows greater impacts among women (46.8%). This situation responds to the crisis of the precarious health system in Mexico and the insufficient budget allocated to it, which together limit the exercise of the right to a healthy life (Benhumea 2019).

Thus, the discrimination experienced in the last five years show that young people are the most affected group in 2017 (24.7%) and 2022 (38.9%), and women have higher values, an increase in this proportion is observed in this period. In 2022, international migrants rank second, with 30.9%, with women again having a higher proportion (35.5%). As can be seen, the percentage of the indigenous population (from 24.0 to 23.1%) and older people (from 17.0 to 15.3%) who reported having experienced at least one discrimination experience decreased from 2017 to 2022 (Fig. 2.6).

Regarding *specific situations of discrimination, during 2022*, the majority of the population from the four analyzed groups reported facing attitudes that affect the exercise of their rights and freedoms on equal terms, specifically encountering various situations where they were made to feel bad or looked at uncomfortably, followed in all cases by receiving insults, being subjected to ridicule, and/or being told things that bothered them. Additionally, in all cases, it is notable that women

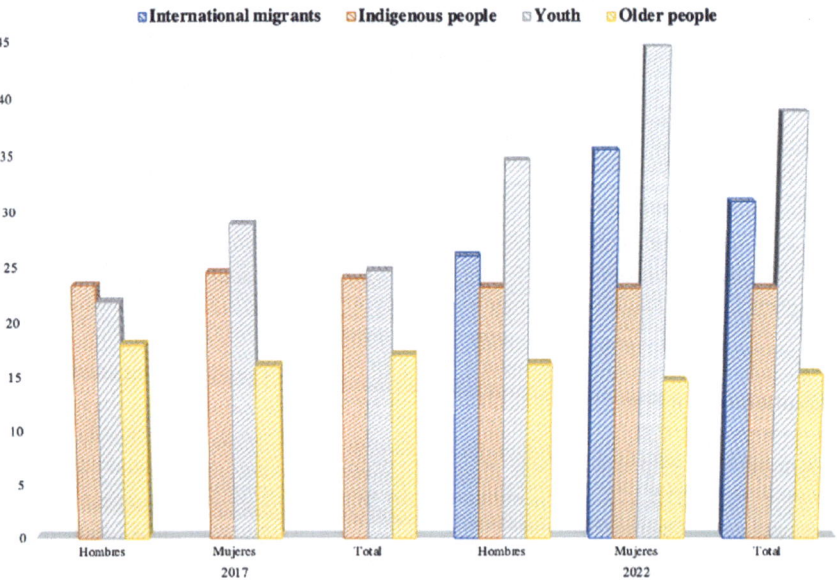

Fig. 2.6 Persistence of discrimination by population group and gender in, the last five years 2017 and 2022. *Source* Own elaboration based on ENADIS (2017, 2022)

2.2 Discrimination Overview: Actors, Practices, and Denial of Rights

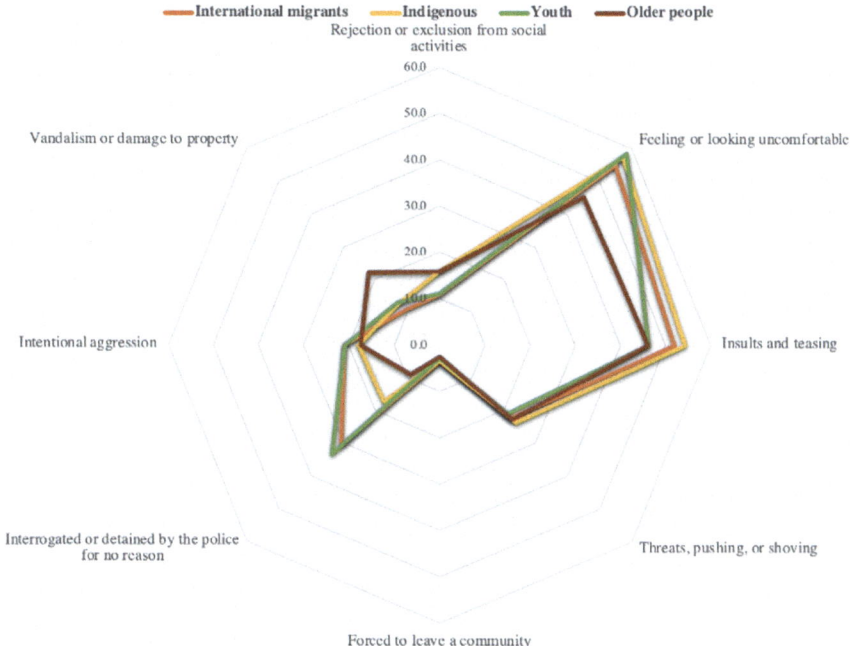

Fig. 2.7 Discrimination situations experienced by population group, last five years 2022. *Source* Own elaboration based on ENADIS (2022)

record higher proportions, which reflects systematic gender-based discrimination (Fig. 2.7).

On the other hand, we draw attention to the discrimination experienced by the migrant population and young people in the face of abuses of authority, particularly when around 30% of them mentioned being questioned or detained without any reason. Once again, these data allow us to talk about unrecognized and invisible racism, which nevertheless materializes in multiple practices of discrimination and denial of rights.

Finally, it is necessary to mention multiple discriminations, in order to make visible the coexistence and concurrence of various reasons why a person is treated less favorably (Lama 2013), for example, as suggested by the data mentioned earlier, for being a woman and a migrant. In this sense, the ENADIS 2022 highlights that, although the majority of people in each group experienced at least one situation of discrimination, there are cases of individuals who experienced two or more situations that put them at a social disadvantage due to prejudices, stereotypes, and stigmas that construct negative images and transcend into practices that violate the human rights of these social groups.

To reinforce the notion of multiple discrimination, we find that the indigenous population is among those who reported being discriminated against to a greater extent for one or more specific reasons in the twelve months before the 2017 survey

(25.3%), a proportion that was 21.7% for young people and 18.3% for the ageing. However, in 2022, it is the young people who report the highest proportion (30.7%), followed by the migrant population (27.8%), indigenous people (26.6%), and the ageing (17.8%). The trend remains in both years and practically for all groups of greater discrimination against women.

In 2022, the reasons for which the migrant population declares to have been discriminated against are for their way of speaking (29.4%), for being a woman or a man (28.5%), for their way of dressing or personal appearance (27.1%), as well as for their political opinions (23.6%) (Fig. 2.8). Of this population, 56.6% indicated having been discriminated against for one of these reasons, while 17.5% specified at least two situations and 25.8% acknowledged three or even up to fifteen reasons as the cause of the discriminations experienced, adding to their condition of being a migrant population.

Among the indigenous population, appearance plays a significant role, as 31.5% of those who reported being discriminated against identified their way of dressing or personal appearance as the main reason, followed by their ethnic background (29.1%), their way of speaking (28.6%), and their weight or height (27.2%) (see Fig. 2.8). In this case, we see that it is a combination of elements that have traditionally been the focus of discrimination and the construction of social prejudices against this

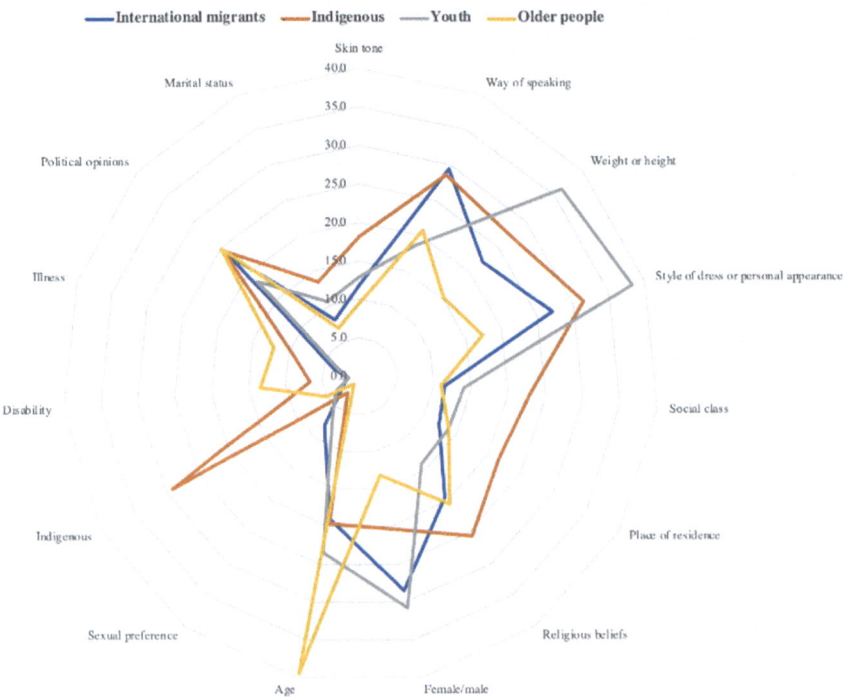

Fig. 2.8 Reasons for discrimination by population group, last twelve months 2022. *Source* Own elaboration based on ENADIS (2022)

2.2 Discrimination Overview: Actors, Practices, and Denial of Rights

population. Regarding multiple discriminations, 40.6% reported being discriminated against for at least one of the aforementioned reasons, while 19.3% identified two reasons, and the rest attributed their experiences of discrimination to more than three elements (40.2%).

As for the young population, the main reason for discrimination is their way of dressing or personal appearance (38.3%), followed by their weight or height (36.3%), gender (30.8%), and age (23.4%). The latter component allows us to recognize that the specific characteristics of each social group, in this case, being young, are among the main causes that reproduce discriminatory practices and attitudes. On the other hand, 44.8% of this population reported being discriminated against for one of the aforementioned reasons, although in this case, the proportion of young people who reported the occurrence of at least two (21.7%) or three situations (12.0%) is higher. In summary, Mexico is a country that disadvantages this population sector.

In the group of older persons, age is the main reason for discrimination (39.4%), followed by those who reported being discriminated against for their political views (24.9%), their way of speaking (20.8%), and their religious beliefs (20.5%) (Fig. 2.8). These situations reveal attitudes that discredit or invalidate the perspectives and ways of thinking of this population.

Among the *main areas of discrimination*, that is, the spaces where these practices occur, differences are observed by groups, as a higher proportion of the migrant population that has been discriminated against identifies social media as the main space (27.5%), followed by government offices (27.0%), streets or public transportation (24.6%), as well as the police or public prosecutor's office (23.9%). This is not the case for the indigenous population, for whom streets or public transportation are the main discriminatory environment (30.7%), as well as work or school (28.4%), and medical services (26.7%) (Fig. 2.9). As can be seen in both cases, it is possible to identify a relationship between denied rights and experiences of discrimination faced by these populations.

Among young people, the three main areas of discrimination are streets or public transportation (41.3%), with a percentage even higher than that recorded by the indigenous population; followed by work or school (35.1%) and social media (22.9%). It is important to note that cyberbullying has become more prevalent, particularly because it involves an environment in which more and more young people are involved, experiencing various risks while being subjected to harassment, humiliation, and other discriminatory expressions.

Finally, among the older people, it is observed that the family, far from being the space that provides mechanisms of protection and support, constitutes the main context in which they experience these situations of discrimination (32.3%), followed by the street or public transportation (31.3%) and health institutions (31.3%). Similar to what happens with multiple discriminations, around 40% of the population in these four groups has been discriminated against in two or more spaces, which expresses the presence of a society for which processes of human rights education and openness to diversity are required, as multiple reasons, contexts, and experiences place these and other populations in a condition of disadvantage and vulnerability.

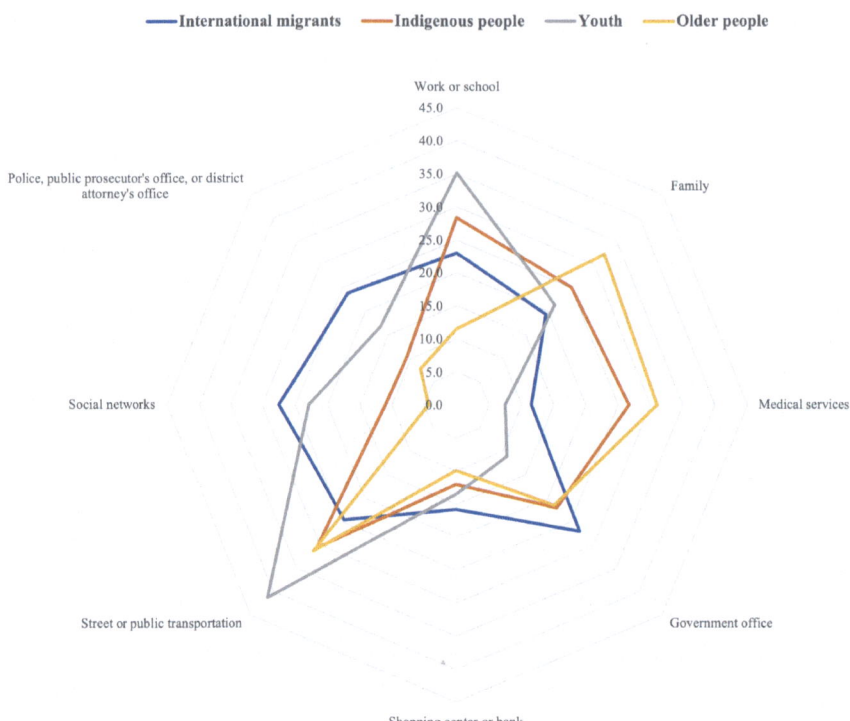

Fig. 2.9 Areas of discrimination by population group, last twelve months 2022. *Source* Own elaboration based on ENADIS (2022)

2.3 Conclusion

As a closing remark, the magnitude of the challenges faced is significant, given that the existence of discriminatory practices in Mexico is widespread and socially very relevant, making their elimination a long-term challenge (Rincón 2005, p. 7). While progress has been made in legal and institutional terms, it is necessary to strengthen these efforts through different strategies, with research and knowledge generation being an indispensable input in the fight against discrimination, as it allows for a deeper understanding of its study and identification of its mechanisms of perpetuation, proposing solutions to the problem (Bucio 2012), and guiding the agenda for equality of treatment.

This also implies joining efforts to transform cultural patterns and eliminate the prejudices that underpin discriminatory practices towards various social groups. Institutional progress is important and constitutes a necessary step, but it is essential to strengthen these advancements by promoting an anti-discriminatory culture, one that not only identifies prejudice and inequality as undesirable but also effectively promotes equal treatment in both public and private institutions, as well as in various aspects of everyday life. Undoubtedly, this is the great challenge for building an egalitarian and discrimination-free society, as discriminatory practices persist despite normative and formal changes, as demonstrated in the contextual exercise presented here.

In other words, statistics on discrimination in Mexico reveal that disdain, along with unjust and unfavorable treatment of vulnerable groups, continues to be a problem that progressively impacts the effective exercise of the rights of these populations. The variables considered in the descriptive analysis provide insight into the multiple manifestations of this social issue among those who perceive themselves and have been discriminated against. Exposing this social reality, questioning the scope of institutional efforts, and highlighting setbacks in this area is part of the work that is prioritized to draw attention to the ongoing need to promote sustainable actions aimed at raising awareness of this issue among Mexican society, its effects on the populations who experience it, as well as the essential awareness and sensitivity required to counteract these practices. The goal is to emphasize that the dialogue between public actions and social and community work must be understood as a joint effort to address this process, which is unnoticed by some, unknown, indifferent, or normalized by others in light of the structural roots of the problem.

The data presented here contribute to the reading and interpretation of the specific cases presented in this book, thus being useful and relevant for the identification of elements that contribute to the design and exercise of evaluation of non-discriminatory public policies.

References

Benhumea, L. (2019). El pacto por México: una reflexión sobre el sistema precario de salud Mexicano. *SAPIENTIAE: Revista de Ciencias Sociais, Humanas e Engenharias*, 5(1), 5–30.

Bucio, R. (2012). Presentación. El desafío de profundizar en el conocimiento de la discriminación para atacar sus raíces. En R. de la Madrid (Coord.), *Reporte sobre la discriminación en México 2012. Introducción general* (pp. 9–11). CIDE, Consejo Nacional para Prevenir la Discriminación.

Cámara de Diputados. (2003). *Ley Federal para Prevenir y Eliminar la Discriminación*. DOF 19-01-2023. https://www.diputados.gob.mx/LeyesBiblio/pdf/LFPED.pdf

Centro de Investigación en Política Pública. (2022). *El desempeño del mercado laboral mexicano: potencia sin aprovechar*. IMCO. https://imco.org.mx/el-desempeno-del-mercado-laboral-mexicano-potencial-sin-aprovechar/#:~:text=En%20total%2C%20el%20potencial%20no,hombres%20tambi%C3%A9n%20tienen%20empleo%20insuficiente

Centro para la Justicia y el Derecho Internacional. (2019, 27 de septiembre). *México: denuncias ante CIDH violación sistemática de derechos humanos contra migrantes* [Comunicado de prensa]. https://cejil.org/comunicado-de-prensa/mexico-denuncian-ante-cidh-violacion-sistematica-de-derechos-humanos-contra-migrantes/

Comisión Nacional para el Desarrollo de los Pueblos Indígenas. (2006). *Percepción de la imagen del indígena en México. Diagnóstico cualitativo y cuantitativo*. CDI. https://www.inpi.gob.mx/2021/dmdocuments/percepcion_imagen_indigena_mexico.pdf

Consejo Nacional para Prevenir la Discriminación. (2018). *Informe Anual de Actividades y Ejercicio Presupuestal 2017*. CONAPRED. https://www.conapred.org.mx/wp-content/uploads/2023/05/InformeAnual2017.pdf

Consejo Nacional para Prevenir la Discriminación. (2019). *Informe Anual de Actividades y Ejercicio Presupuestal 201.8*. CONAPRED. https://www.conapred.org.mx/wp-content/uploads/2023/05/InformeAnual2019.pdf

Consejo Nacional para Prevenir la Discriminación. (2020). *Informe Anual de Actividades y Ejercicio Presupuestal 2019*. CONAPRED. https://www.conapred.org.mx/wp-content/uploads/2023/05/InfomeAnual2018.pdf

Consejo Nacional para Prevenir la Discriminación. (2021). *Informe Anual de Actividades y Ejercicio Presupuestal 2020*. CONAPRED. https://www.conapred.org.mx/wp-content/uploads/2023/05/InformeAnual2021.pdf

Consejo Nacional para Prevenir la Discriminación. (2022). *Informe Anual de Actividades y Ejercicio Presupuestal 2021*. CONAPRED. https://www.conapred.org.mx/wp-content/uploads/2023/05/InformeAnual2021.pdf

Consejo Nacional para Prevenir la Discriminación. (2023). *Informe Anual de Actividades y Ejercicio Presupuestal 2022*. CONAPRED. https://www.conapred.org.mx/wp-content/uploads/2023/06/InformeAnual2022.pdf

Consejo Nacional para Prevenir la Discriminación. (n.d.a). *Informes anuales del CONAPRED 2004–2020*. CONAPRED. https://www.conapred.org.mx/transparencia/planes-programas-e-informes/informes-anuales-del-conapred-2004-2020/

Consejo Nacional para Prevenir la Discriminación. (n.d.b). *Programa Nacional para la Igualdad y No Discriminación (PRONAIND)*. CONAPRED. https://www.conapred.org.mx/pronaind/

Consejo para Prevenir y Eliminar la Discriminación de la Ciudad de México. (2021). *COPRED llama a visibilizar las brechas de género persistentes en el Día Internacional de la Mujer Indígena*. COPRED. https://www.copred.cdmx.gob.mx/comunicacion/nota/copred-llama-visibilizar-las-brechas-de-genero-persistentes-en-el-dia-internacional-de-la-mujer-indigena

Félix, C., Spijker, J. y Zueras, P. (2023). Sistema de pensiones y apoyo social a adultos mayores en México 1979–2019. *Papeles de Población, 27*(110), 79–107.

Francioli, S. & North, M. (2021). Youngism: The content, causes, and consequences of prejudices toward younger adults. *Journal of Experimental Psychology General, 150*(12), 1–22. https://doi.org/10.1037/xge0001064

Gracia, M. A y Horbath, J. E. (2019). Condiciones de vida y discriminación a indígenas en Mérida, Yucatán, México. *Estudios Sociológicos, 37*(110), 277–307. https://estudiossociologicos.colmex.mx/index.php/es/article/view/1666/1786

Gobierno de México (2025). Pensión para el bienestar de las personas adultas mayores https://programasparaelbienestar.gob.mx/pension-bienestar-adultos-mayores/

Heatley, A. (2021). Jóvenes y desigualdad en México: ¿el derecho de piso de una sociedad adultocéntrica? *Intersticios sociales*, (21), 71–98. https://www.intersticiossociales.com/index.php/is/article/view/305/569

Instituto Mexicano de la Juventud. (2017). *¿Qué es ser joven?* Gobierno de México. https://www.gob.mx/imjuve/articulos/que-es-ser-joven

Instituto Nacional de Estadística y Geografía. (2022). *Encuesta Nacional sobre Discriminación (ENADIS) 2022*. INEGI. https://www.inegi.org.mx/programas/enadis/2022/

Instituto Nacional de Estadística y Geografía. (2023). *Para Saber +. Encuesta Nacional sobre discriminación*. INEGI. https://www.inegi.org.mx/contenidos/programas/enadis/2022/doc/enadis2022_infografia.pdf

Lama, A. (2013). *Discriminación múltiple* (tomo LXVI). ADC. https://revistas.mjusticia.gob.es/index.php/ADC/article/view/3715/3715

Naciones Unidas. (2023). *Temas. Relator especial sobre el derecho a la salud*. NU. https://www.ohchr.org/es/special-procedures/sr-health

Pérez, G. y Oliver, P. (2020). *Migraciones, derechos humanos y comunicación intercultural en los ambientes de trabajo. Cartilla de sensibilización*. Organización Internacional para las Migraciones. https://www.r4v.info/es/document/cartilla-de-sensibilizacion-migraciones-derechos-humanos-y-comunicacion-intercultural-en

Raphael, R. (Coord.) (2012). *Reporte sobre la discriminación en México 2012. Introducción general.* CIDE, Consejo Nacional para Prevenir la Discriminación.

Rincón, G. (2004). Presentación. En J. Rodríguez Zepeda, *Qué es la discriminación y cómo combatirla* (pp. 5–6). CONAPRED.

Rincón, G. (2005). Rasgos y retos de la lucha contra la discriminación en México. *El Cotidiano, 21*(134), 7–11.

Rodríguez, J. (2018). Sensatez y sensibilidad: cómo construir una institución antidiscriminatoria en un país fragmentado. En Consejo Nacional para Prevenir la Discriminación (Coord.), *Por la igualdad somos mucho más que dos. 15 años de lucha contra la discriminación en México* (pp. 49–64). Secretaría de Gobernación, CONAPRED.

Rodríguez, J. (2004). *Qué es la discriminación y cómo combatirla.* CONAPRED. Rodríguez, J. (2018). Sensatez y sensibilidad: cómo construir una institución antidiscriminatoria en un país fragmentado. En Consejo Nacional para Prevenir la Discriminación (Coord.), *Por la igualdad somos mucho más que dos. 15 años de lucha contra la discriminación en México* (pp. 49–64). Secretaría de Gobernación, CONAPRED.

Secretaría de Gobernación. (2003, 11 de junio). *DECRETO por medio del cual se expide la Ley Federal para Prevenir y Eliminar la Discriminación.* DOF: 11/06/2003. https://www.dof.gob.mx/nota_detalle.php?codigo=694195&fecha=11/06/2003#gsc.tab=0

Secretaría de Gobernación. (2011, 10 de junio). *DECRETO por el que se modifica la denominación del Capítulo I del Título Primero y reforma diversos artículos de la Constitución Política de los Estados Unidos Mexicanos.* DOF:10/06/2011. https://dof.gob.mx/nota_detalle.php?codigo=5194486&fecha=10/06/2011#gsc.tab=0

Secretaría de Gobernación. (2021). *Programa Nacional para la Igualdad y No Discriminación 2021–2024.* CONAPRED. http://www.conapred.org.mx/wp-content/uploads/2023/05/PRONAIND_2021-2024.pdf

Chapter 3
"The Mexicos as Perceived and Lived": An Approach to the Discrimination Experiences of Populations in Human Mobility Contexts

The diverse and myriad economic, social, political, and environmental factors shaping the "backbone of migration" (Guillén et al. 2019, p. 283) are among the causes of the increased presence of mobile populations: from 221.0 to 280.6 million from 2010 to 2020, a figure that represents 3.6% of the world's population in the last year (BBVA Mexico and CONAPO 2023). On the international stage, Mexico ranks second among the top ten countries of origin for migrant populations (11.2 million), with the Mexico-United States migration corridor being the most dynamic globally (3.9% of global migration in 2020) (BBVA Mexico and CONAPO 2023).

In addition to these dynamics, this country faces institutional and budgetary challenges in responding to this multidimensional phenomenon, considering that "it has gone from being a territory almost exclusively of emigration to becoming one through which all kinds of migrants and individuals in need of international protection transit and settle" (Sánchez and Zedillo 2022, p. 4). In this regard, these authors warn that Mexico's migration context represents one of the most complex in the world, due to the dynamism of migration flows, the diversity of profiles and needs of these populations, as well as the countless instances of violence and governmental management incapacity (González and Aikin 2023); all explanatory elements of discrimination (Gutiérrez et al. 2020).

In this sense, discrimination among populations in mobility contexts takes on other nuances and levels of complexity, as Mexico's migration policy "still focuses on controlling human mobility and restricting irregular migration, and lacks articulated and coherent actions to promote full integration and the exercise of human rights without discrimination" (Sánchez and Zedillo 2022, p. 16). This situation undermines individuals who move "without papers" (Ramos 2016), as the notion of "illegality" and the criminalized representation of these individuals reproduce discourses, stigmas, stereotypes, prejudices, and xenophobic positions that often materialize in practices that violate and limit the exercise of human rights (Estrada et al. 2022; Varela et al. 2021).

© The Author(s), under exclusive license to Springer Nature Switzerland AG 2025
A. E. Jardón Hernández et al., *Multiple Discriminations*,
SpringerBriefs in Environment, Security, Development and Peace,
https://doi.org/10.1007/978-3-031-85826-0_3

Therefore, the objective of this chapter is to analyze the processes of discrimination against populations in situations of human mobility in and by the State of Mexico. Through the presentation of selected cases, an approach to the complexity of migration dynamics in non-border contexts is offered, where the presence of refugee populations, transit migrants, and returnees also demands the active participation and intervention of states, as suggested by Rodríguez (2023), based on structural strategies grounded in a human rights approach that contributes to the construction of peaceful and inclusive societies.

The chapter includes a first section that prioritizes actions promoted within the framework of the SDGs and the Global Compact for Safe, Orderly, and Regular Migration (GCM) in Mexico, specifically to counteract discrimination against mobile populations. In a second section, the cases are described through narratives that capture the experiences of individuals in migration and mobility contexts.

3.1 Actions for Building Peaceful, Inclusive, and Non-discriminatory Societies

The social and cultural roots of discrimination outlined above allow for an understanding of the relationship between migration and mobility phenomena and public discourses of hate that dehumanize, criminalize, and discriminate against these individuals. In other words, the concepts and images constructed about migrant persons "shape society's opinion and attitude" (OXFAM México 2023, p. 15). In this regard, perceptions gathered in the study conducted by this organization show, for example, that conversations about migration on social media platforms reproduce such discourses and recreate images associated with aporophobia -as discussed earlier- justifying rejection attitudes based on the poverty situation they face.

> The United States relies on Mexico to deal with dangerous migrants trying to reach North America.
>
> **Migrants, due to their economic need**, are more likely to end up involved with drug trafficking and cartels.
>
> **Only poor migrants come to Mexico**, who possess lesser social and economic value for society.
>
> Returned Mexican migrants are unfamiliar with the country and its culture, and their opinion about it should not be taken into account. (OXFAM México 2023, p. 25)

Hence, the relevance of universal calls such as the SDGs, proposed with the purpose of promoting peace and access to justice, among others, for historically discriminated populations becomes evident. Within the framework of the fourth SDG, quality education, specifically in target 4.7, it is pertinent to note that human rights education requires a comprehensive perspective based on education for global citizenship and education for the appreciation of diversity—all dimensions of this target, to make it a key tool for managing processes of human mobility and reducing practices of discrimination, racism, and xenophobia experienced by these individuals in Mexico (Jardón et al. in press). At a more specific level, articulation with SDG 16 is also considered, as the protection of the rights of these populations is based on the application of laws and policies aimed at promoting respectful interaction among different cultures, customs, and traditions; that is, in the construction of inclusive, tolerant, peaceful, and secure societies.

In this regard, the work of the Mexican State has sought to materialize within the guidelines of the United Nations Framework for Sustainable Development Cooperation in Mexico 2020–2025 through actions supported by an intersectional perspective.

> The Mexico Cooperation Framework concluded its third year of implementation in 2022. Over these three years, the United Nations has worked to support the state in addressing income inequalities, living conditions, educational levels, economic and sociocultural factors, security, transparency, access to health services, and the exercise of rights, which are exacerbated by discrimination and vulnerabilities stemming from individuals' characteristics, such as gender, age, belonging to indigenous peoples or Afro-descendant populations, sexual orientation, disabilities, nationality, legal status, or place of residence, among others. (NU México 2022, p. 20)

This tool encompasses an integrated approach to leaving no one behind through five programmatic principles: Human Rights, Gender, Interculturality, Life Cycle, and Territory (NU México 2022, p. 20). Among these, human rights and interculturality are key in implementing strategies focused on migrant and refugee populations. According to the four areas of work outlined in this Cooperation Framework, two stand out for their direct relationship with the study population: (i) Equality and Inclusion and (ii) Peace, Justice, and Rule of Law. These areas recorded the highest number of initiatives developed in 2022.

As an example, in the area of equality and inclusion, it is reported that around 783 government institutions, civil society organizations, academia, and the private sector strengthened their capacities, among other aspects, to promote interculturality and/or combat discrimination. In more specific terms, assistance was provided to 15,650 individuals in their local integration processes, with 4561 refugees successfully placed in formal employment (p. 30). These actions, as part of a social, economic, and labor inclusion program, contribute to advancing the protection of the right to work, which, as indicated in the previous chapter, is among the main rights violated among this population.

In the area of Peace, Justice, and Rule of Law, the Program for Legal Orientation and Assistance for Persons with International Protection Needs in Mexico is reported. The main objective is to provide information, advice, and legal representation to asylum seekers, refugees, and others in situations of human mobility, to make them aware of their options, rights, and obligations (p. 54). This is strategic considering that misinformation is one of the main barriers to accessing protection guarantees. While the various actions promoted impact different SDGs, it is worth noting that, of the 244 initiatives counted, 13.9% of these (34) correspond to SDG 16, making it one of the objectives that received the most attention and budgetary allocation in that year.

On the other hand, regarding the GCM, the argument of this work aligns with objective 17, which aims to eliminate all forms of discrimination and promote evidence-based public discourse to change perceptions of migration. This is crucial in the face of hate speech, criminality, and various stereotypes and prejudices towards mobile populations.

Therefore, with the adoption of the GCM, the Mexican government committed to monitoring the 23 objectives of this compact, proposing thereby to implement a new approach to migration policy focused on respecting human rights and addressing the structural causes of migration. According to the UPMRIP (2022a), the actions implemented by various national and state agencies and institutions in Mexico have focused their efforts on reducing vulnerabilities, access to services, access to information, and eliminating all forms of discrimination. As can be seen in Table 3.1, initiatives focused on the latter theme are among those with the highest number of programs aimed at changing negative perceptions and attitudes toward migration.

In summary, monitoring the SDGs and the GCM regarding the link between migration and discrimination highlights that it is a topic of interest on the government agenda. However, it is also noted that the structural connotation of this issue requires continuity in work and solid strategies that address the complexity of the problem, with the aim that visible results in this area are achieved in the medium and long term.

Table 3.1 GCM objectives with the highest number of programs and/or actions, 2022

Objective		Number of programs
7.	Addressing and reducing vulnerabilities in migration	46
15.	Providing migrants access to basic services	33
3.	Providing accurate and timely information at all stages of migration	27
4.	Ensuring all migrants have proof of their legal identity and appropriate documentation	25
1.	Collecting and using accurate and disaggregated data for policy formulation	23
17.	Eliminating all forms of discrimination and promoting evidence-based discourse to change perceptions of migration.	24

Source UPMRIP (2022b, p. 8)

3.2 Discrimination in the Experience of Mobility and Migration

Considering that experiences of discrimination among migrant populations are a constant in various spheres and spaces of life, this section offers an approach to specific cases that shed light on this issue among refugee populations, transit migrants, and Mexican returnees to the State of Mexico. As it was marked in the introduction of this book, the cases presented in this chapter are drawn from different research projects on migration, mobility, and labor, from which various interviews were conducted to identify multiple discriminatory situations and practices in the migratory experience of these individuals. Within the framework of these research projects, these populations were interviewed at different times linked to migratory dynamics and observed national and state-level circumstances, such as the increased registration of asylum applications from the Venezuelan population fleeing their country due to the crisis experienced there, the processes of collective organization of so-called migrant caravans, and the increase in the number of Mexicans returned by US immigration authorities.

The cases presented here do not aim to generalize but to contribute to the understanding of this issue in non-border geographical spaces, where the complexity reached in this and other issues has not received sufficient attention in academic and governmental agendas. The importance of focusing on the State of Mexico responds, among other factors, to the proximity and/or vicinity that several of its municipalities have with Mexico City, as they tend to become a concentrator "node" of the population seeking refugee status in the city but looking for work and housing in the conurbation area of the state, where they can rent spaces to live at lower prices and access different job offers (personal communication with the founder of the INTRARE organization, 2022). On the other hand, the presence of transit migrants is increasingly visible, particularly in the municipalities that are part of the route or journey of "La Bestia," the notorious "train of death" that represents various risks and dangers for many of these migrants when attempting to board it to reach the northern part of the country and cross into the United States (Medellín 2024).

Briefly, in figures, the modalities of flows and populations in mobility considered in this work record the following dynamics:

- Mexico ranks among the countries with the highest number of asylum seekers. In 2021—following the COVID-19 health emergency—the number of asylum applications reached a record high of over 130,000 (ACNUR 2023, p. 18), with 2023 marking the year with the highest number of these: out of the 140,982 applications, the majority came from Haiti (44,239), Honduras (41,935), Cuba (18,386), El Salvador (6117), and Guatemala (6111). Additionally, among the nine offices and representations that the Mexican Commission for Refugees Assistance (COMAR) has in different entities of the country, Mexico City ranked second in terms of the number of applications recorded until December 2023 (30,872) (COMAR 2023). This dynamic has an impact on the

municipalities of the conurbation area of the State of Mexico, given the housing and employment options sought by this population in these spaces (Gandini et al. 2021).
- On the other hand, an approach to the population of transit migrants is seen in the registration of foreign nationals in irregular situations. According to the UPMRIP, the events corresponding to this population have significantly increased over the last 15 years, rising from 120,445 to 444,439 from 2017 to 2022. Of these, in the last year, 69.4% are men and 30.6% women, with the majority of them coming from Venezuela (21.8%), Honduras (16.4%), Guatemala (15.6%), Cuba (9.3%), and Nicaragua (9.2%) (BBVA and CONAPO 2023, p. 93). Out of the 32 federative entities, the State of Mexico ranks 14th concerning the events of individuals in irregular situations (5348) (UPMRIP 2022b).
- Finally, regarding the return of Mexican population, this chapter focuses on the repatriation events from the United States, considering that it is the flow that tends to be criminalized and discriminated against to a greater extent, due to the conditions under which their return occurred, as well as the sanctions with which some of them return and which prevent them from entering that country for some years and/or for life. In this regard, according to data from the UPMRIP (2022b), out of the 258,000 events recorded in 2022, approximately 12,171 correspond to population of Mexican origin, a figure with which this entity ranks among the top eight that register the highest presence of these events.

Therefore, the complexity of the migration issue in the State of Mexico, and the dynamism, and volume of various international migration flows (Ramírez 2025), necessitate focusing attention on this entity where discriminatory practices transcend into multiple violations of human rights for this vulnerable group.

3.2.1 Escaping the Crisis: In Search of Refugee Status and Family Reunification

The Martinez family's experience began with the migration of Jhoel, who left his native Venezuela in 2016, anticipating the economic collapse and humanitarian crisis that worsened in the country in later years. Jhoel entered Mexico as a tourist but with the expectation of starting a new life in this country with his wife, María, and their three children. They agreed that the family would travel to Mexico once he could afford the expenses.

Supported by his social networks, Jhoel decides to travel to the municipality of Toluca, where he would receive guidance from his friend Génesis, a Venezuelan who has been living in this country for several years. With Génesis's support, he secures his first job at a retail store chain, where he was paid minimum wage due to lack of work documents. Consequently, the possibility of obtaining a job commensurate with his skills and professional profile becomes complicated by his subsequent irregular migration status, with misinformation and lack of awareness,

3.2 Discrimination in the Experience of Mobility and Migration

as previously mentioned, being some of the main risks hindering the exercise of rights while reproducing discriminatory practices (Organización Internacional para las Migraciones [OIM] 2015).

Therefore, his migratory status, being a foreigner and of Venezuelan origin, were the main reasons Jhoel considers having faced multiple difficulties in accessing a job where he wouldn't be exploited and would receive a sufficient salary to cover his needs and have the means to bring his family. Hence, discriminatory practices in the workplace by employers, coworkers, and other Mexicans have been the main issue Jhoel has faced, as well as the main obstacle to realizing his family reunification project.

However, his insistence on finding a better job leads him to receive support from an acquaintance responsible for an insurance agency, where he is initially hired without documents and later seeks to regularize his situation through a work visa for foreigners, although this requires him to leave the country for a consular interview at a Mexican embassy. And so he did, he took the risk, and traveled to Guatemala to fulfill this requirement, but the visa was denied by the National Institute of Migration (INM) on suspicion that the application was being made for migratory purposes rather than strictly labor-related.

After this experience, he decided to return to Mexico, and this re-entry is what he says allowed him to submit his asylum application retroactively in August 2017 when he found that possibility while researching online and initiated the process from the INM representation office in the State of Mexico, where the lack of knowledge on the part of the public servants did not make the process easy, nor did it allow it to proceed expeditiously.

In this scenario, the months that Jhoel projected to reunite his family turned into approximately two years, among other factors, due to the denial of his right to work and human dignity. The possibility of planning his family's trip materialized through the support of what he considers "church brothers" of the Church of Jesus Christ of Latter-day Saints [Mormons], who provided him with the money to buy tickets for his wife and three children.

By then [late 2017], the Venezuelan government significantly restricted the exit of its population, so it was not easy to "evade the authorities" and leave the country. The family's anguish and desperation of "fleeing their country" were compounded by the deportation experience that María and her children endured, as Mexican immigration authorities rejected their entry and returned them to Bogotá after withholding their documents and subjecting them to psychological mistreatment. This situation is attributed to Jhoel's irregular status, but it does not justify the criminalization and dehumanization they faced in their attempt to enter Mexico, nor does it justify the violation of their right to not be detained incommunicado and to be treated equally under the circumstances.

> It's like they make you feel you're a criminal, that's how they treated me. When I entered, they told me, "Your husband is not legal here, and until he is, you cannot enter." I was terrified. All I did was cry. They kept me waiting for hours in a room, and then they passed me to another person who just told me to write my husband's name on a little white paper they gave me. So, I did it. I didn't dare to lie or put another name, even though I had money

to enter, even though I had my invitation letter, even though I had everything. They didn't ask me for anything else, just his name. Then, on the 9 a.m. flight, they sent me back. But they accompany you with someone from the airport who guides you to the plane, and there they also take away your documentation, they take your passport. The same person, when we arrived, took me to the ticket counter in Bogotá, where they handed over the documentation. They asked me if I wanted to continue because my flight was from Cúcuta to Bogotá and then from Bogotá to Mexico. They asked me if I was going back to Cúcuta, and I said no, they were leaving me here. I collected my bags, and when I made that decision, they gave me back my passport, and I stayed in Bogotá. (María, 46-year-old Venezuelan, San Antonio la Isla 2022)

Without being able to communicate with her husband, María and her children arrived in Bogotá. Aware that she couldn't return to Venezuela, that she had already "done the hardest thing, leaving the country," she chose not to continue her journey back to Cúcuta. Once she managed to establish communication with her husband, he asked María to travel to Ecuador to stay with his brother-in-law's family. Out of dissatisfaction and anger over what happened to her family, Jhoel went to the offices of the CNDH to file a complaint. During this visit, he learned about the legal assistance and advice provided by the organization Sin Fronteras, IAP to support and advocate for the rights of people in human mobility.

Through contact established with this organization, they began receiving a series of advisories focused on ensuring the right to family reunification. They received guidance on the discourse and attitude María should adopt before migration authorities and were urged to immediately notify her intention to seek asylum, by the principle of non-refoulement enshrined in Article 33 of the 1951 Convention.

The security Jhoel gained through these advisories led the family to plan their departure from Ecuador to reunite in Mexico, which became possible in November 2018. However, on this second occasion, authorization for their entry was not easy. Despite being advised and psychologically prepared for possible threats and treatment by immigration authorities, María was again detained and held incommunicado for six hours because she had already been notimobilefied from Bogotá that she was coming to seek asylum in Mexico.

Fearing that his family would be deported again, Jhoel made sure to stay in touch with personnel from Sin Fronteras and the Federal Attorney's Office for the Protection of Children and Adolescents, even taking the initiative to notify them of his family's arrival at INM offices in the State of Mexico. The strategies adopted by Jhoel made it possible to form a commission with representatives from the CNDH and the Attorney's Office, which had to follow up on the case and go to the airport to prevent deportation, given the situation of detention and violation of María and her children's rights. This ultimately led to authorization for entry into Mexico and the reunion of this family.

When the family entered Mexico, Jhoel had not yet received approval for refugee status, so to ensure their legal status to stay in this place, they followed the recommendation not to link María and her children's asylum application to his, as both requests could be rejected, and Jhoel's process had already progressed. It was in January 2019 that he was notified of the approval of his application. To date,

María and her children have permanent visas, but unlike Jhoel, they do not have refugee status, as the process was done by requesting a family unit visa.

The reading of this case highlights how social differentiation by nationality, economic status, and gender reveals structural and intersectional discrimination processes, as well as the violation of multiple rights, including the right to non-discrimination, the right not to be criminalized, the right to family unity, the right to human dignity, the right not to be detained incommunicado, the right not to be detained, and the right to the best interests of the child.

This example also reflects on the intersection of xenophobia and aporophobia as elements that transcend in the exercise of these discriminatory practices, both due to the rejection of people from other nationalities and the disdain and mistreatment María and her children received due to the complex economic situation facing their country.

In conclusion, Jhoel's recounting of his family's experience reminds of "the Mexicos that are experienced and perceived: the one that embraces you and the one where authorities treat you like a criminal." This underscores the need to advance and strengthen education on rights to acquire tools, skills, and intercultural competencies aimed at eradicating these practices that violate these vulnerable groups.

3.2.2 Cultural Vulnerability Among Transient Migrant Populations

Another migratory modality that receives attention in this chapter is transit migration, as it involves flows that became more visible with the deployment of so-called migrant caravans from Central America. The start of these displacements occurred in October 2018 when a caravan was called for in Honduras through social media under the slogan "We are not leaving because we want to: we are expelled by violence and poverty." A week later, 160 people gathered in San Pedro Sula, and by October 17, around four thousand people had congregated at the border bridge with Mexico with the intention of crossing and reaching the United States (El Colegio de la Frontera Norte 2019).

The transit of this population through Mexico entails a multitude of risks, abuses, and violence that are exacerbated by the involvement of organized crime (Kuhner 2011). Additionally, intolerance and lack of respect for the customs and traditions of foreign populations in Mexican society are among the main challenges they face, leading to accusations and instances of discrimination, as happened in the well-known case of *#ladyfrijoles*, who expressed her displeasure with the food provided in a shelter, saying: "the food they are giving us is awful [...] all ground-up beans [...] as if they were feeding pigs." This comment was enough to trigger the disdain of a large number of Mexicans who questioned what she ate in Honduras and demanded her return to her country, labeling them as freeloaders and ungrateful (Pérez and Aguilar 2021).

According to Bustamante (2002), the vulnerability in which the foreign population is placed is due, among other factors, to the characteristics of structural discrimination, as it represents a form of social differentiation and domination where power relations and schemes of domination/subordination produce various vulnerabilities, including cultural vulnerability arising from anti-immigrant ideologies, xenophobia, and racism (Fig. 3.1).

Within the framework of these displacements, the State of Mexico became a territory of transit and temporary rest for migrants who could stay in one of the two shelters officially established in the municipalities of Apaxco and Huehuetoca (BBVA Research 2020). The narratives that are highlighted to expose the situations of discrimination and vulnerability faced by migrants in transit correspond to individuals who were interviewed at an improvised shelter located in the municipality of Metepec, founded by a civil society member, and advocate for the human rights of these populations.

Thus, the asymmetry and abuse of power are observed in lived experiences that generate fears and acts of resistance among this population, as was the case with

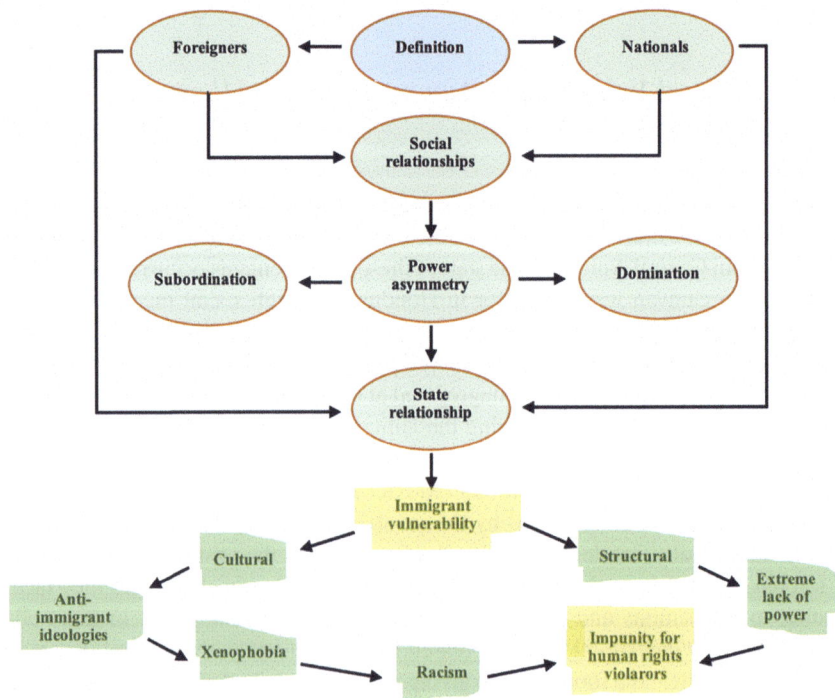

Fig. 3.1 Vulnerability of the migrant population. *Source* Own elaboration based on Bustamante (2002, p. 341)

3.2 Discrimination in the Experience of Mobility and Migration

Carlos, who during his stay at the shelter had a confrontation with individuals apparently from the ministerial police. According to his narrative, they came to the shelter premises intending to take away the women and children, which Carlos prevented, although this triggered subsequent acts of aggression, violence, and confiscation of belongings. The gravity of these events is further compounded by the institutional abandonment expressed by Carlos, as in his attempt to report the incidents, he was ignored and discriminated against due to his nationality and migratory status in the country.

> When I was here [at the Metepec Shelter] as a caretaker, I defended the other fellow companions, well, the minors. The first time, three ministerial officers came in, they put a gun to our heads because they wanted to take out the women and girls, the children, and I stood in their way and told them that if they had the guts, they could shoot, but I wasn't going to move, that they weren't going to take anyone, and so, I pushed them out until they left. Then, at around one in the morning, they came back, pushing the gate, trying to force it open, the ministerial officers, and well, they also took things from the guys who had come from work, their cell phones, even their shoes, socks, they pulled down their pants in front of the women and children, there they stripped them to take their money, and they threatened to come back to kill me. I reported it, but they never did anything, I made many reports and Human Rights did nothing because they told me to go back to where I came from. These people from Toluca have never helped us. (Carlos, 35-year-old Salvadoran, Metepec, 2021)

Nevertheless, the exposure to risks and dangers is a topic that this population has become aware of, as expressed by Maria when she says that as women they are exposed to "anything happening to them", or that due to cultural constructs towards the foreign population they face situations of discrimination and other acts of contempt.

> Honestly, I never imagined how my experience would be, but I knew what I was risking because you know that when you leave your country or your home, you know what you're risking. I knew I was risking anything. One, because I'm a woman, and two, because you know many people are very selfish, and racist, and, for example, when I arrived here, nobody wanted me. (María, 34-year-old, Honduran, Metepec, 2021)

Therefore, the rejection of foreigners, migrants, and those in poverty, as experienced by Laura and Carlos, goes beyond discrimination, generating situations of violence and helplessness in protecting their human rights.

> Let's say that when you're out on the street asking for money, people can be mean, really rude. Sometimes they yell really ugly things at you, and sometimes they tell us to go back, go back to your country. 'What the hell are you doing here?' (Laura, 35-year-old Honduran, Metepec, 2021)

> When the pandemic started, I used to greet people and they would say, 'No, no, no, don't touch me. Damn migrant, you're the ones who bring the damn coronavirus'. I would just say, 'Yeah, no problem'. I stopped going to places I liked, they started rejecting me, and well, I hardly show up anymore. These are situations where they blame us and we haven't done anything. (César, 35-year-old, Salvadoran, Metepec, 2021)

A revealing finding in terms of these processes of social differentiation by specific nationalities is found in the results of the ENADIS 2022, specifically in the stance that

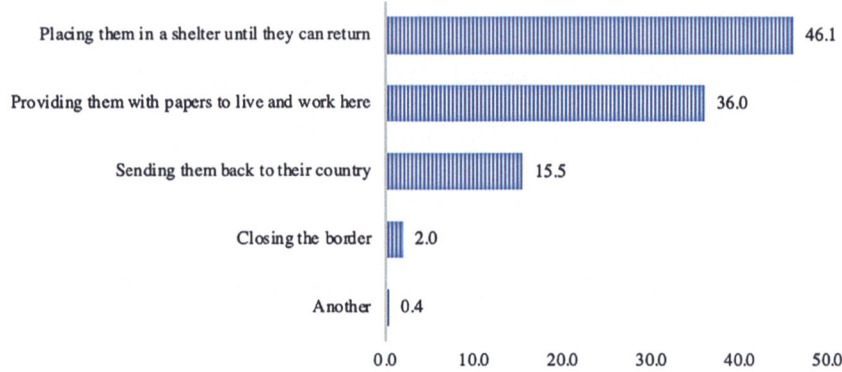

Fig. 3.2 Stance on the Mexican government's actions regarding the Central American population. *Source* Own elaboration based on ENADIS, 2022

the surveyed population aged 18 and over has regarding what the government "should do" with Central American people arriving in Mexico. Among the response options to this question, the stance that leans towards tolerance, inclusion, and acceptance is "granting them papers so they can live and work here" (36.0%) (Fig. 3.2).

On the other hand, the attitudes that reinforce discrimination and xenophobia are those of individuals who believe that the government should place them in temporary shelters until they can return to their country (46.1%), followed by those who think they should be sent back to their birthplaces (15.5%), and those who suggest closing the borders to them (2.0%). In other words, approximately 63.6% of the surveyed population holds negative attitudes that promote rejection and exclusion of this population (Fig. 3.2).

The differentiation of these attitudes at the state level reveals that in the central region (State of Mexico and Mexico City), there is a higher proportion of surveyed individuals with a negative stance towards this population (Fig. 3.3). These findings suggest the absence of conditions conducive to peaceful coexistence under tolerance frameworks that promote acceptance and inclusion.

3.2.3 From Stigma to Denial of Rights for Deported Population from the State of Mexico

In 2007, amidst the debate on the new phase in Mexico-United States migration, various theses emerged associated with the observed changes. Among these, the emphasis on return migration found support in the increased number of Mexicans returning from the United States: from 670 to 2,940 thousand between the periods of 1995–2000 and 2005–2010 (Passel et al. 2012). These returns, driven by the international economic crisis as well as Barack Obama's anti-immigrant policies,

3.2 Discrimination in the Experience of Mobility and Migration

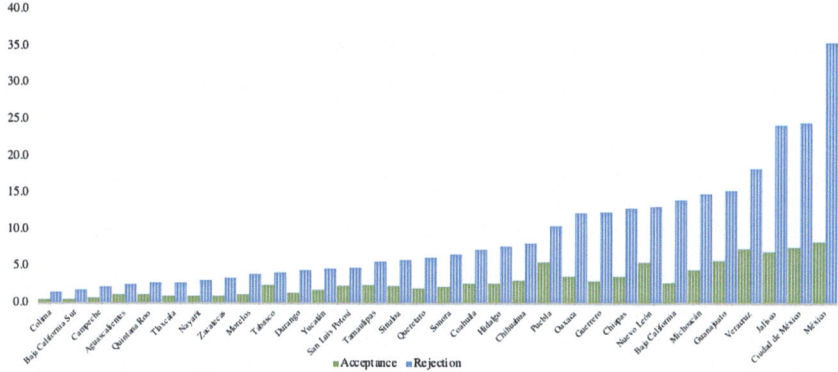

Fig. 3.3 Attitudes of rejection and acceptance towards Central American Population by State. *Source* Own elaboration based on ENADIS, 2022

included both deported individuals and those who chose to return voluntarily out of fear, job loss, and other reasons.

In this scenario, the predictions of an imminent and massive return led to the media portrayal of the issue (Alarcón et al. 2009; Centro Latinoamericano y Caribeño de Demografía [CELADE] 2010; Cruz and Zapata 2013) reinforcing prejudices and stereotypes towards the population returning under various circumstances, particularly those with several years of residence in the United States.

> Mexico continues to experience migrants returning, just in the State of Mexico alone last year in two thousand nineteen we had over ten thousand deportees from the United States. So, if you add to that the ones being deported from the United States and those attempting to cross without resources, it becomes a problem for us because where are we going to get the money? Or how are we going to provide comprehensive care? You understand it as a public official, and you have to say, 'Well, I will do everything within my power', but when they are requesting the process, it's like, 'I want the support now'. So, it's a very important and impactful issue of both political and economic strength for the country because the message should be one of kindness towards this population (Interview with a state official 2020).

Based on these elements, this section explores the experience of the population from the State of Mexico who were deported and have returned to the municipality of Tenancingo. The presentation of the narratives aims to emphasize the prejudices towards this population, as well as the implications that discriminatory practices have on denying the right to decent work and access to social programs.

The cultural motives of discrimination, that is, prejudices and stereotypes (Rodríguez 2023), can be identified in cases like Sergio's, who was deported in 2018 after spending around 15 years in the United States. In addition to the long period of absence, which in itself tends to complicate reintegration into the life and origin spaces of this population, attitudes of aversion or hostility further complicate their processes of adaptation and integration (Majidi 2020; Schuster and Majidi 2013). Sergio's narrative highlights that belonging to "the group of the deported" implies, in some cases, the assignment of certain labels -lazy, criminal, failure- that put them at risk of discrimination or lead to its occurrence.

> I'm not ashamed to say that I'm deported, but it's not something I want to go around saying either, because it's been difficult for me to be here. I picked up a really bad habit in the United States: alcoholism. Sometimes you hear it, and other times we just know they're saying that we couldn't make it in the United States, that we were deported for being lazy, that we don't want to work here anymore, or even that they blame us for the insecurity: us, the deported ones. (Sergio, 37-year-old, Mexican, Tenancingo, 2020)

When discrimination occurs, that is, when discrimination materializes, multiple rights are violated. The cases of Magali and Claudia, both deported in 2019, highlight another difficulty faced by this population: rivalry and competition as relationships that produce conflicts and reduce even the possibilities of self-employment. The narratives of Magali and Claudia allow us to understand at least four elements. First, the complexity of reintegration both in the processes of emigration and in return, as it is a continuum where the risk of discrimination is present because it involves a population that "struggles here and there." As a second element, the reproduction of discourses that place the returned migrant as a population that comes to take away jobs, that is, from this perspective, it seems that they are individuals without rights and needs, so, in practice, as Goffman (2006) refers, different types of discrimination lead to reducing their life chances. Thirdly, the need to promote and reinforce education in human rights to raise awareness and sensitize different sectors of society about the rights of this population and other vulnerable groups. Finally, the unequal treatment and contempt towards this group of people are based on various prejudices that, in addition to placing them in a situation of social disadvantage, have significant impacts on the protection of their rights and access to opportunities.

> Finding a job when you return is not easy, but you shouldn't give up, especially if you keep telling yourself, 'No, it's difficult, no, it's difficult'. In reality, it has always been that I arrived in the United States and struggled, I arrived in Mexico and also struggled, but it's because, as a fellow worker here in Tenancingo says, 'they want to take our jobs away from us', I mean, ¡how is that possible if I'm from here! When I returned, I continued with my rebozo work, and people didn't want to buy from me because it's difficult for them after all those years, I mean, there's a lot of selfishness and rivalry here in Mexico, not everyone supports you, not everyone says, 'come, I'll teach you'. So, after being in the United States and returning, it becomes even more difficult to reintegrate into a job because of this rivalry issue, because they see us as competition, because people think we don't need it because we made money up north or because they say we have bad habits because we were deported. (Magalí, 46-year-old, Mexican, Tenancingo, 2020)

> I learned to make pastries in the United States, but I was embarrassed to go to bakeries to ask if they needed people because some acquaintances said that after so many years in the United States, I couldn't start my own business. Also, I think there is a lot of envy towards us [those with migration experience]. For example, some people sell desserts; there was a time when my mother-in-law and I were selling dinner, but then a lady set up on the corner, about halfway down the block, and sold the same thing as us. But the lady would even stand up and say, 'What did they sell you? How much did they give them to you for?' People would tell us, 'We don't come to buy from you anymore because the lady on the corner stops us and asks: what are they? How much are they?' So we said, "No more, let's not sell anymore." (Claudia, 31-year-old, Mexican, Tenancingo, 2020)

To reinforce this notion of unequal treatment and its impacts on the opportunities available to this population, in the narratives of Magalí and Claudia, it is possible to interpret—according to the data from the ENADIS 2022 presented in chapter two—that the main violated right among this population is access to social programs, with a proportion that is higher among women. Government offices and their officials are the main areas and actors of discrimination, as is the case in these instances, where the condition of being a woman, a migrant, deported, and with other sociocultural experiences, places them at a disadvantage compared to others.

> When one approaches government offices to request support, there are indeed differences. For example, I requested support once for sewing machines because I knew they had supported several people who were skilled in sewing. They told me, 'Let's submit applications for sewing machines'. I submitted applications with a councilman, but he never submitted the paperwork. Whenever I asked him about the progress of my machines, he would say, 'I haven't received a response yet'. I could see that other people were receiving support. So, one time, I got angry and told him, 'Either they haven't responded, or you didn't submit those papers'. That was because his secretary told me, 'Ma'am, I'll be honest with you. Here are the papers; the councilman hasn't submitted them' (Magalí).

> I also requested support for some covers for the plants I sow [mesh for ornamental plants], and it never arrived even though I submitted it three times. So, they treat us differently and deny us support, and I think that's discrimination. They always said they would contact me, but I never heard anything. I went to the town hall several times to inquire, and it was always the same: 'We'll call you', but they never did (Claudia).

Altogether, the narratives of the deported population shed light on the problems and difficulties that arise from both discrimination processes and the denial of rights among those who have been forcibly returned, particularly in the absence of concrete public strategies and actions that facilitate their reintegration and equal integration into the economic, social, political, and cultural life of the places to which they are returning.

3.3 Conclusion

The analysis of these cases contributes to the understanding of structural and intersectional discrimination, as the accumulation of disadvantages among this population affects the full enjoyment of their human rights. Moreover, social differentiation processes based on specific conditions place some individuals in more vulnerable situations than others. Finally, from the narratives that capture the voices, perceptions, and experiences of the subjects involved in this research, it is clear that the institutional actions and efforts implemented so far are not only insufficient but also limiting given the complexity of mobility and migration processes.

Therefore, proposing that education promoting tolerance and diversity is sufficient to address the right to non-discrimination ignores the fact that humans are predisposed to prejudice and intolerance. Thus, addressing the structural foundations of this issue requires:

> Identifying the *practices* that give meaning to racist discourse involves highlighting the actors and spaces where the *doxa* (stereotypes and prejudices) is reproduced. This includes examining representations *of self* and *otherness* in the media, as well as beliefs of superiority/inferiority that underpin discriminatory discourse against migrants. These discourses not only legitimize social inequality but also hinder the exercise of their rights during their transit through national territory. (Varela et al. 2021, p. 13)

In other words, structural changes aimed at combating discrimination and racism require active promotion and implementation of priority policies to reduce inequalities. Commitment, sustainability, and budget allocation are key elements to move beyond rhetoric and good intentions, to ensure effective access and protection of the human rights of people in human mobility. So far, progress has been insufficient, as migrants in transit through Mexico are victims of multiple crimes that are rarely reported, investigated, or punished, in addition to the fact that migrant detention centers are overcrowded, unsanitary, and dangerous, and the asylum system is severely overwhelmed (Human Rights Watch 2024). The outlook for Mexican migrants returned from the United States is also not encouraging, as during the recently outgoing administration of Andrés Manuel López Obrador (2018–2024), there was reduced attention to this population group, due to cuts in support for migrants in general, and in particular to return programs (Jacobo and Cárdenas 2021). Among other factors, this relates to the insufficient priority given to human mobility in the government's public agenda and budget, as well as the entrenched schemes of administrative corruption that limit the scope of public initiatives, while also impacting and restricting the work of civil society and other human rights defenders of this population.

The recognition of the population in human mobility as rights-bearing subjects is posed as one of the main challenges to progressively overcoming the stereotypes, prejudices, and discriminatory practices that fuel both the reduced institutional response to these processes, as well as the rejection and exclusion that stem from insufficient awareness among authorities and society at large. "Hilos que nos unen" (Threads that unite us) is positioned as a new anti-discrimination and anti-xenophobia campaign in Mexico aimed at promoting positive behaviors towards migrants.

Nevertheless, this type of initiative created by international organizations such as the IOM lacks visibility in territories like the State of Mexico, given the centralization of their actions in cities with large migrant populations. Recognizing the role that the State of Mexico plays in migration and human mobility is crucial to strengthening the promotion of these activities in its various municipalities. Specifically, to make progress in this area in the State of Mexico, it is necessary to begin by identifying, recognizing, and making visible the dynamism and diversity of the various migratory flows that converge in its territory and interact with its population. This will open up spaces for coordinated work needed to achieve the goals of the GCM and the SDGs.

References

Agencia de la ONU para los Refugiados. (2023). *Caminando hacia la integración. Principales resultados*. ACNUR México. https://www.acnur.org/mx/media/caminando-hacia-la-integracion-2022-principales-resultados-de-acnur-mexico

Alarcón, R., Cruz, R., Díaz, A., González, G., Izquierdo, A., Yrizar, G. y Zenteno, R. (2009). La crisis financiera en Estados Unidos y su impacto en la migración mexicana. *Migraciones Internacionales*, 5(16), 193–210. https://migracionesinternacionales.colef.mx/index.php/migracionesinternacionales/article/view/1108

BBVA México y CONAPO. (2023). *Anuario de Migración y Remesas*. https://www.bbvaresearch.com/wp-content/uploads/2024/03/Anuario_Migracion_y_Remesas_2023.pdf

BBVA Research. (2020). *Mapa de casas del migrante, albergues y comedores en las principales rutas de migración por México*. https://www.bbvaresearch.com/wp-content/uploads/2020/02/Mapa_2020_Albergues_Migrantes_Portable.pdf

Bustamante, J. (2002). Immigrants' Vulnerability as Subjects of Human Rights. *International Migration Review*, 36(2), 333–354. http://www.jstor.org/stable/4149456

Centro Latinoamericano y Caribeño de Demografía. (2010). *Impactos de la crisis económica en la migración y el desarrollo. Respuestas de política y programas en Iberoamérica*. CELADE. https://kmhub.iom.int/sites/default/files/impactos_de_la_crisis_economica_en_la_migracion_y_el_desarrollo_0.pdf

Comisión Mexicana de Ayuda a Refugiados. (2023). *La COMAR en números*. COMAR. https://www.gob.mx/comar/articulos/la-comar-en-numeros-327441?idiom=es

Cruz, R. y Zapata, R. (2013). Naturalización y vulnerabilidad de los inmigrantes mexicanos en Estados Unidos. En M. E. Anguiano y R. Cruz (Eds.), *Migraciones internacionales, crisis y vulnerabilidades. Perspectivas comparadas*. El Colegio de la Frontera Norte.

El Colegio de la Frontera Norte. (2019). *Caravanas migrantes de Centroamérica*. El Colef. https://www.colef.mx/estemes/caravanas-migrantes-de-centroamericanos/

Estrada, J.P., Ávila, M. J., y Martínez, M. L. (2022). La discriminación histórica a personas migrantes en tiempos de la pandemia de la COVID-19 en Coahuila, México. *Huellas de la Migración*, 7(13), 11–43. https://doi.org/10.36677/hmigracion.v7i13.16595

Gandini, L., Fernández, A., Narváez, J.C., Rodríguez, L.H., Franco, M., Pilatowsky, E. y Rojas, R. (2021). *Documento de trabajo 1. Protección social de las personas refugiadas y solicitantes de la condición de refugio en México. Un análisis de oportunidades y capacidades institucionales*. Organización Internacional del Trabajo. https://www.ilo.org/wcmsp5/groups/public/---americas/---ro-lima/---ilo-mexico/documents/publication/wcms_838086.pdf

Goffman, E. (2006). *Estigma. La identidad deteriorada*. Amorrortu.

González, A. y Aikin, O. (2023). (In)Movilidad humana en México en contextos de vulnerabilidad, crisis regionales y políticas de cierre de fronteras. *Análisis Plural* (3), 1–17. https://doi.org/10.31391/ap.vi3.45

Guillén, J. C., Menéndez, F. G., y Moreira, T. K. (2019). Migración: Como fenómeno social vulnerable y salvaguarda de los derechos humanos. *Revista de Ciencias Sociales (Ve)*, XXV(E-1), 281–294. https://doi.org/10.31876/rcs.v25i1.29619

Gutiérrez, J.M., Romero, J., Arias, S.R., y Briones, X.F. (2020). Migración: Contexto, impacto y desafío. Una reflexión teórica. *Revista de Ciencias Sociales (Ve)*, XXVI(2), 299–313. https://doi.org/10.31876/rcs.v26i2.32443

Human Rights Watch. (2024). *Informe mundial 2024. Migrantes y solicitantes de asilo*. HRW. https://www.hrw.org/es/world-report/2024/country-chapters/mexico

Jacobo, M. y Cárdenas, N. (2021). Back on your own: migración de retorno y la respuesta del gobierno federal en México. *Migraciones internacionales*, 11 https://doi.org/10.33679/rmi.v1i1.1731

Jardón, A., López, A. y Martínez, N. (en prensa). Educación en Derechos Humanos. Claves para contrarrestar la xenofobia hacia la población en movilidad humana. En L. Delgadillo (Coord.), *Debates, desafíos y propuestas sobre vulnerabilidad e inclusión social*. UAEMéx, UNESCO.

Kuhner, G. (2011). La violencia a las mujeres migrantes en tránsito por México. *Opinión y debate*, (6). https://corteidh.or.cr/tablas/r26820.pdf

Majidi, N. (2020). Assuming reintegration, experiencing dislocation – returns from Europe to Afghanistan. *International Migration*, 59(2), 186–201. https://doi.org/10.1111/imig.12786

Medellín, C. (2024). Huehuetoca el camino al norte, ruta de la esperanza para muchas familias de migrantes. *La Silla Rota*. https://lasillarota.com/metropoli/2023/9/23/huehuetoca-el-camino-al-norte-ruta-de-la-esperanza-para-muchas-familias-de-migrantes-449086.html

Naciones Unidas México. (2022). *Informe de resultados 2022. Trabajando en conjunto para cumplir la promesa de no dejar a nadie atrás. Oficina de Coordinación Residente del Sistema de Naciones Unidas en México*. NU. https://mexico.un.org/es/232966-informe-de-resultados-2022

Organización Internacional para las Migraciones. (2015, 14 de julio). *Comunicado Global. La OIM apoya campaña de información en México para proteger a los migrantes* [Comunicado de prensa]. https://www.iom.int/es/news/la-oim-apoya-campana-de-informacion-en-mexico-para-proteger-los-migrantes

OXFAM México. (2023). *El muro mexicano. Estudio de percepción sobre la migración en México*. OXFAM México. https://oxfammexico.org/wp-content/uploads/2023/08/EMM_Informe_completoR4.pdf

Passel, J., D'Vera, C. & González-Barrera, A. (2012). Net Migration from Mexico Falls to Zero-and Perhaps Less. Pew Hispanic Center. https://www.pewresearch.org/race-and-ethnicity/2012/04/23/net-migration-from-mexico-falls-to-zero-and-perhaps-less/

Pérez, M. y Aguilar, M. (2021). #LadyFrijoles: señalamiento, discriminación y estigma de migrantes centroamericanos a través de redes sociales en México. *Andamios*, 18(45), 223–243. https://doi.org/10.29092/uacm.v18i45.817

Ramírez, T. (2025). Cambios y continuidades en los patrones migratorios y movilidades poblacionales en el Estado de México. En A. Jardón (Coord.), *Escenarios de las movilidades y migraciones contemporáneas en el Estado de México*. Universidad Autónoma del Estado de México. https://ri.uaemex.mx/handle/20.500.11799/141973

Ramos, M. (2016). *Reconocimiento, derechos humanos e intervención social. Migrantes en el noreste de México*. México. Universidad Autónoma de Nuevo León.

Rodríguez, J. (2023). *Una teoría de la discriminación*. Universidad Autónoma Metropolitana-Iztapalapa.

Sánchez, E. y Zedillo, R. (2022). La complejidad del fenómeno migratorio en México y sus desafíos. *Serie de Documento de Política Pública. Elementos para entender los retos de la migración* (pp. 3–33). PNUD América Latina y El Caribe. https://www.undp.org/es/latin-america/publicaciones/la-complejidad-del-fenomeno-migratorio-en-mexico-y-sus-desafios

Schuster, L. & Majidi, N. (2013). What happens post-deportation? The experience of deported Afghans. *Migration Studies*, 1(2), 221–240. https://doi.org/10.1093/migration/mns011

Unidad de Política Migratoria, Registro e Identidad de Personas. (2022a). *Informe ejecutivo 2022. Pacto Mundial para una migración Segura, Ordenada y Regular en México*. UPMRIP https://portales.segob.gob.mx/work/models/PoliticaMigratoria/Documentos/Informe_PMM_2022.pdf

Unidad de Política Migratoria, Registro e Identidad de Personas. (2022b). *Boletines Estadísticos. Personas en situación migratoria irregular*. UPMRIP. http://www.politicamigratoria.gob.mx/es/PoliticaMigratoria/CuadrosBOLETIN?Anual=2022&Secc=3

Varela, A., Ruíz, V. y Pech, C. (2021). Racismo, migración y discriminación. El trabajo de la Re-presentación. *Andamios*, 18(45), 9–20. https://doi.org/10.29092/uacm.v18i45.808

Chapter 4
Discrimination Against Indigenous Women from the State of Mexico Who Work as Domestic Workers in Mexico City

As outlined in the contextual chapter, in Mexico, Indigenous people face discriminatory attitudes in various social spheres such as work, school, and healthcare services. As a culturally ingrained and socially widespread behavior of contempt (Rodríguez 2023), discrimination against Indigenous people in Mexico is rooted in racism and prejudices associated with ethnic origin.

Thus, being an Indigenous person represents a disadvantage in Mexico not inherently due to their ethnicity but because they are embedded in a society that perpetuates racist behaviors, where ethnic identity is neither appreciated nor socially valued. Prejudices associating Indigenous people with backwardness, ignorance, servitude, and poverty restrict their social mobility and full access to non-discrimination. As Gutiérrez argues, racism "contributes to perpetuating stereotypes and racist prejudices, as well as activating a repertoire of symbolic violence expressed through mockery and ridicule, thereby signaling limitations for individuals to transcend their condition of poverty, marginalization, and discrimination" (2015, p. 124).

The experiences of discrimination faced by Indigenous people are not homogeneous, thus necessitating an intersectional analysis. As discussed in the theoretical chapter, intersectionality allows for the identification of various groups subjected to different forms of discrimination detrimental to their human rights and dignified treatment, giving rise to various dimensions of disadvantage. As Gracia and Horbath (2019) point out, intersectionality is a fruitful approach for analyzing the discrimination faced by Indigenous people in urban environments, as it considers how stigmatizing constructions associated with ethnic identity combine and amplify with discrimination based on social class and gender (p. 282).

Building on the above, the objective of this chapter is to analyze the discrimination faced by Indigenous women working as domestic workers[1] in

---

[1] In this chapter, the term "domestic worker" is used, as the International Labour Organization (ILO) does. Nevertheless, it should be noted that activists in Mexico have advocated for the use of another

Mexico City. Specifically, emphasis is placed on the conditions of Mazahua Indigenous women from the State of Mexico, who have a historical labor migration to Mexico City. The analysis will focus on domestic workers who work under the "live-in" arrangement, meaning they perform cleaning and sometimes caregiving tasks and reside in their employers' homes, receiving a weekly wage, housing, meals, and one day off per week (Durin 2017). As explained in the chapter, Indigenous women domestic workers face discrimination due to their ethnic origin, gender, and the nature of their work, which is often associated with servitude. Those living in their employers' homes are subjected not only to long working hours but also to confinement and other abuses that violate their dignity and rights.

The analysis presented here focuses on the labor aspect, as already outlined in the contextual chapter, the right to work is one of the main rights violated for the populations under analysis in this book. In the case of domestic workers, their work is performed in a privately considered sphere, in activities that are undervalued and result in a lack of dignified treatment for Indigenous women domestic workers due to multiple discriminations that are often overlooked.

The chapter is organized into four sections. The first section presents some considerations about the labor migration processes of the Mazahua people to Mexico City, the main destination for work outside the State of Mexico, with special emphasis on the discrimination processes in the urban context and the labor integration of Mazahua women as domestic workers. Subsequently, a brief exposition is made about the working conditions of domestic workers in Mexico, which are characterized by their precariousness and facing conditions of discrimination.

The third section conducts an analysis of the multiple discriminations faced by indigenous women domestic workers who work under the "live-in" arrangement. In this section, emphasis is placed on the experience of Mazahua women from the State of Mexico, based on literature on Mazahua migration and information gathered in fieldwork. The chapter closes with a section that emphasizes the importance of public actions to dignify domestic work in Mexico, and the need to promote cultural changes to prevent stigmatization and discrimination processes towards indigenous women domestic workers.

The analysis presented here is based on academic research and studies conducted by CONAPRED on indigenous domestic workers, the diagnosis carried out by the COPRED with the Government of Mexico City on domestic workers, as well as testimonies of women who have acted as activists in favor of labor rights and social recognition of this population sector. Information about the Mazahua ethnicity is derived from a review of specialized literature on Mazahua migration to Mexico City, a historical destination for this ethnic group, with particular emphasis on mentions about domestic work. Likewise, some experiences of women from the Mazahua

---

term in Spanish language, namely "home worker" (*trabajadoras del hogar*) to avoid the notion of domesticity, linked to servants and contemptuous expressions. Even if it is recognized for the case of Mexico, and in order to follow the expression commonly used in English language, along this chapter it is used the term "domestic worker".

region of the State of Mexico who have worked as domestic workers in Mexico City are recovered from academic literature; in addition to the information gathered in interviews conducted in 2022 in San Felipe del Progreso and Jocotitlán.

4.1 Mazahua Migration to Mexico City and Labor Niches

Since the 1940s, within the framework of intense rural-urban migration, Mexico City, the capital of the Mexican Republic, has been a destination for indigenous migrants from various ethnic groups, originating from different states of the country in search of work and better living conditions. The urban space of Mexico City, while favoring labor integration and earning monetary income, has also become a space of discrimination against indigenous people.

This discrimination is based on representations that stigmatize indigenous-peasant identities and associate them with negative attributes, and there is a lack of public policy actions that consider the specificities of this population, which joins the city life under disadvantaged conditions (Gracia and Horbath 2019). Thus, indigenous migrants to Mexico City face difficulties in accessing basic social rights such as decent housing, public health, and well-paid jobs.

The labor activities in which they are inserted are characterized by their precariousness, low wages, and lack of labor rights, such as construction, street vending, and domestic work (Sánchez 2004, p. 73). Although these are not the only occupations for indigenous people, they still maintain a significant presence as activities to be developed in Mexico City and reflect the challenges faced by the indigenous population to improve their working conditions over time.

Indigenous groups from the State of Mexico have also participated in migratory flows to Mexico City. The State of Mexico stands out for the remarkable dynamism of internal migration to various entities of the Mexican Republic in search of work, with Mexico City as the main destination (Ramírez 2025). The rural communities of the Mazahua region in the State of Mexico, located in the municipalities of San Felipe del Progreso, San José del Rincón, Atlacomulco, Ixtlahuaca, and Jocotitlán (Fig. 4.1), maintain an active dynamic of labor mobility to different entities, and as at the state level, Mexico City is the main destination for work (Hernández and Jardón 2018; Hernández 2025).

It is a historic destination for Mazahua labor migration, in which settlement processes have been reported (Oehmichen 2005), while significant circularity is maintained, facilitated by road infrastructure, the availability of public transportation, and relative proximity due to the border between Mexico City and the State of Mexico. This allows people working in Mexico City to return to their communities every week or every fifteen days, where they maintain their permanent residence (Hernández 2025).

Arizpe (1978) points out that starting from the 1940s, Mexico City begins to be the main destination for temporary Mazahua migration, supported by community networks that provide information, lodging support, and job placement. Since then,

Fig. 4.1 Municipalities of origin of Mazahua migration to Mexico City. *Source* Elaborated by Luz María Ledesma Reyes based on Hernández, 2025

their employment has been reported in precarious jobs clearly differentiated by gender: for men, work in masonry at various points in the city, as well as porters, "macheteros," and employees in the La Merced market area and other establishments.

On the other hand, women have worked as street vendors of fruit or crafts in public spaces wearing their Mazahua attire, and have been derogatorily referred to as "Marias." They have been discriminated against for their appearance in public spaces and various establishments, as they do not receive the same treatment as other people due to their appearance. Additionally, due to their commercial activity, they have also faced violence from city authorities, who mistreat them in public institutions, confiscate their merchandise, and have even been detained (Oehmichen 2005). In addition to their work as street vendors, another job niche for Mazahua women has been domestic work (Oehmichen 2005; Arizpe 1978), commonly known as "household work".

Faced with the lack of job opportunities in their communities of origin, domestic work has been an option for Mazahua women, usually young, allowing them to contribute financially to their households. Due to wage differences, relative proximity, and support from family and community networks, the main destination for household work has been Mexico City. This work for women has persisted for decades, as reported in studies by Arizpe (1978), who noted that young women

4.2 Brief Considerations on Domestic Work in Mexico

from the Mazahua region, "unless they have a special reason to stay in their region of origin, all go to Mexico City" (p. 125).

Domestic work as an option for Mazahua women has been reported in the municipalities of San Felipe del Progreso (Arizpe 1978; Oehmichen 2005), Atlacomulco (López 2017), and San José del Rincón (González 2015). It is also very common for young women to work on a live-in basis to save on accommodation and food expenses, but as will be discussed later, this also lends itself to extended working hours and labor exploitation.

Although there seems to be a connection between the concepts of "Marias," "servants," and "Mazahuas" (Ojarasca 2005), domestic work in Mexico City is not solely an occupation for Mazahua women but for indigenous women in general. In fact, there seems to be a social imaginary connection between indigenous women and domestic work or cleaning work (Gutiérrez 2015). In a context of poverty and scarce job and educational opportunities in their communities of origin, young and single indigenous women find domestic work as a labor insertion option in the city. As discussed below, this is precarious employment, "where they face situations of abuse, exploitation, and discrimination that reproduce situations of poverty and exclusion" (CONAPRED 2008, p. 118).

4.2 Brief Considerations on Domestic Work in Mexico

Prior to analyzing the discrimination against indigenous women domestic workers, it is worth noting some particularities of this work, which is characterized by its invisibility, lack of recognition, and labor regulation. Firstly, it is essential to highlight that it is a feminized job, as the majority of domestic workers are women. Additionally, it is precarious work, often performed without written contracts, access to social security and labor benefits, and with wages below the minimum established by law. This reflects a lack of recognition of the formal employment relationship between employers and domestic workers as rights-bearing individuals (COPRED et al. 2021, p. 5).

Domestic work is considered "help" associated with servility and activities traditionally assigned to women due to their gender. The work performed by domestic workers is deemed socially and economically inferior, and as previously mentioned, they do not receive social benefits. Since most workers obtain employment through friends, family, or existing employers, formal employment contracts are rare (Bautista 2012). This fosters ambiguity regarding working hours and specific job duties, "allowing discrimination, exploitation, abuse, and mistreatment" (CONAPRED 2011, p. 3). In this sense, the treatment domestic workers receive depends on the goodwill of their employers and is not established as part of their labor rights and dignified treatment.

This undervaluation of domestic work is also reflected in the way women domestic workers have been labeled (Bautista 2022). Terms like "maid", "servant", "cleaning lady" or "domestica" are commonly used. Employers (mostly women), on the other hand, are referred to as "madam," and their workers designate them as "bosses." These distinctions serve "the daily restoration of unequal limits, that is, it is part of the daily exercise of power" (CONAPRED 2008, p. 37), and as Durin (2017) suggests, these terms encapsulate views on otherness, hierarchies, and inequalities. This worker-employer relationship is framed within a dynamic of power and domination (Gramsci 1975), to the extent that employers complain that their workers "talk back" to them, as a transgression of the boundaries established by this hierarchical relationship (Goldsmith 1981).

As González (2012) argues, the precariousness defining domestic work has a dual dimension: it is socially undervalued work, lacking labor rights, and subject to ridicule and contempt towards the women who perform it. Furthermore, domestic work "is another way of reducing women to the position of reproducers of the domestic order, facilitators of family care, and confined to the private sphere, which is supposed to be untouchable by social integration policies" (p. 92).

Studies on domestic work indicate the greater vulnerability of workers who work under the "live-in" arrangement, meaning they reside in their employers' homes. While this allows workers to save on accommodation and meal expenses, it has been reported that the rooms allocated for their rest are in poor condition or lack privacy, there are restrictions on the use of various spaces or utensils, the assignment of leftover or poor-quality food, practices that reinforce the social distance between workers and employers (Durin 2017; González 2012).

Particularly, in live-in work, working hours are often exhausting, as they can extend to more than 12 h. Likewise, there is no distinction between living space and workspace, so workers are susceptible to confinement and greater control over them. This results in emotional exhaustion and stress, which are reported elements for domestic workers in Mexico City, and to a lesser extent, instances of violence such as cuts, unjustified severe reprimands, shouting, burns, preventable falls, mistreatment, violence, harassment, confiscation of belongings, and sexual violence (COPRED et al. 2021, p. 53).

4.3 Indigenous Women Working as Domestic Workers: Multiple Discriminations

In Mexico City, efforts have been made to raise awareness about the condition of domestic workers. In September 2020, the "Survey on the Rights Situation of Domestic Workers in Mexico City" was launched and received 428 responses, which were collected through the website of the COPRED. It is worth noting that the call banner was translated into Mazahua and Mixe, with videos in both languages (COPRED et al. 2021). The results of this survey highlight the precariousness of this work, the lack of labor rights, and the discrimination to which they are subjected.

Among the results, it stands out that 31% of the domestic workers surveyed consider that they have been discriminated against for some reason in their work. Within this group that has suffered discrimination, the reasons were their economic condition (38.2%), physical appearance (30.9%), and social status (25%); ethnic origin (8.1%), language (5.9%), clothing style (15.4%), and gender (8.1%). In response to this, it is striking that the majority of people (74.3%) did nothing, which largely responds to the normalization of discriminatory acts, lack of trust in institutions where the credibility of the complaint may be questioned, fear of retaliation from the employer, and/or losing their job, as well as ignorance of the institutions that can address these cases (COPRED et al. 2021).

Regarding the salary received, the report indicates that 24.5% of the domestic workers surveyed consider that non-indigenous domestic workers receive a higher salary than indigenous domestic workers. Faced with these perceptions, the report allows us to confirm "that almost 60% of the domestic workers who belong to or identify with any indigenous people or community earn less than a minimum wage or a minimum wage" (COPRED et al. 2021, p. 46).

From an intersectional perspective, the analysis of discrimination situations highlights how axes of social inequality such as gender, social class, and ethnic belonging affect the treatment that domestic workers receive from their employers, who are generally women of different ethnic origins and social classes. "This not only generates a labor relationship but also a bond of superiority-inferiority towards the domestic worker" (CONAPRED 2015, p. 6).

The treatment received by indigenous women who work as domestic workers is related to the devaluation of the work they perform, which is considered women's work and of little relevance (Bautista 2012), but also to discrimination based on their ethnic origin and social class, as they are generally women from impoverished rural communities with no other job alternatives. In labor terms, they are susceptible "to exploitation, mistreatment, and even workplace and sexual harassment by their employers" (Centro Nacional para la Capacitación Profesional y Liderazgo de las Empleadas del Hogar A.C. n.d. p. 6), which is often endured out of fear of losing their jobs and income.

Those who are newcomers, although they may have a higher level of education than their predecessors 40 years ago, also face unfamiliarity with the customs, manners, consumption, and way of life of their employers, particularly when they come from a very affluent background, placing them at a disadvantage. They must learn to use appliances that they did not have in their home communities and to prepare and consume foods that were not part of their diet. In some cases, the fact that they are still learning restricts their ability to negotiate with their employers, where the activities to be performed are not clearly defined (Durin 2017).

This is linked to a view of indigenous people as submissive, ignorant, with low education and preparation, "less worldly than urban people," and to whom a favor is done by giving them employment as if they were not deserving of dignified treatment for being indigenous. Authors like Durin (2017) highlight the preference of some employers for young "country" women who are considered more docile and obedient.

Likewise, it is assumed that they will accept the working conditions imposed on them because they need to work (Gutiérrez 2012).

There is, therefore, a devaluation of the work they perform, because they are women, and because of their social and ethnic background. Thus, as pointed out in a CONAPRED study on the social treatment of indigenous women working as domestic workers, "Gender, class, and ethnic discrimination shape serious situations of labor exploitation under the veil of 'service,' and in this sense, it can be affirmed that indigenous women domestic workers experience triple discrimination daily" (CONAPRED 2008, p. 26).

The situations described here coincide with those reported by Mazahua women who have worked as domestic workers in Mexico City. López (2017), in her analysis of Mazahua women working as domestic workers between 1960 and 1970, reports that employers did not like women to speak their language, and even prohibited it as a form of control in a subordinate relationship. Similarly, there were reports of mistreatment in learning processes at work, such as the use of appliances that women did not know how to use because they did not have them in their home community. Likewise, they had to acquire other habits regarding their appearance, such as how to style their hair and dress (Oehmichen 2005).

As already mentioned, live-in work entails greater vulnerability to situations of discrimination and lack of dignified treatment. This violation of dignified treatment translates for indigenous women not only into the prohibition of using their language but also into the spaces and utensils designated for them, a matter that is not only exclusive to Mazahua women but also to indigenous workers, as Lorenza Gutiérrez, activist, recounts:

> in this sense, when we work live-in, we go through many things. We are forbidden to speak in our language because the employer thinks we are speaking ill of her. Exclusive utensils are provided for us, as we are not allowed to use the tableware cutlery, or glasses of our employers. We will also have an assigned place in the kitchen to eat, and we will do so after the family has finished eating. (Gutiérrez n.d.)

Given that this is work carried out in the domestic sphere and the everyday living space of the employers, and therefore under their rules, it is possible to find logics of dominance such as isolation, exploitation, control of movement, and racialization. In this sense, the interviewed women reported "humiliations" (ill-treatment) as part of their experiences in domestic work, due to situations of confinement or even lack of food when they worked in Mexico City:

> They would belittle me, the employers, who then would humiliate you terribly, or not even feed you. (Daniela, 31-year-old, Mexican, Jocotitlán, 2022)

Similarly, the long hours associated with live-in domestic work are a constant theme in narratives about household work, where it is expected that the worker "be there" attending to the needs of their employers, as indigenous women's domestic work is associated with servitude-like activities.

> Yes, I stay there, I'm living there [in the house where she works], well, it's tough because practically one gets up... not so early, but seven, seven-thirty. The good thing is that there's

only the couple, and by eight or nine at the latest, I would rest. But you have to be there. And there are jobs where you start at six in the morning and you don't rest until ten, eleven at night. Yes, because I've been through several jobs and I know. (Angie, 36-year-old, Mexican, San Felipe del Progreso, 2022)

Another aspect is the asymmetrical and paternalistic nature between employers and workers, leading to their infantilization, which makes it difficult to negotiate salary increases, for example. In that sense, it would seem that, to receive good treatment, it depends on finding employers who understand their situation as indigenous workers as if it were a matter of luck and not a matter of exercising rights (Sánchez et al. 2004). The treatment these women receive is linked not only to the working conditions they face (without contracts and benefits) but also to the performance of undervalued activities and their indigenous status.

Gender stereotypes and the notion of servitude associated with indigenous people come into play, seemingly justifying treatment that places them in conditions of greater vulnerability, exclusion, and discrimination. Thus, "the articulation of these inequalities is what allows them to be seen as people susceptible to exploitation or being put under guardianship, or as workers with few resources and many needs" (Durin 2017, p. 29).

Thus, domestic work, although it allows for valuable income for the reproduction of their households in their places of origin, is a job whose conditions are endured out of necessity. In that sense, it is an undesirable job for their daughters:

Do you know why I made sure all my daughters got an education? Because I was humiliated once, in the city, and I said to myself: my daughters - because I have four daughters and a son - my daughters will not be servants like me. I took charge of that. Yes, because I used to tell my daughters: that some people take advantage. I used to have to get up at six in the morning and go to bed at one in the morning to work, I was still ironing at midnight. Yes, and they only gave us one hour [for lunch]: so eat quickly because we have to go do other things; and I said: oh, I endure it for my daughters. (Lucero, 47-year-old, Mexican, San Felipe del Progreso, 2022)

These situations, although illustrated in the case of Mazahua women, are shared by the majority of indigenous live-in domestic workers. As mentioned earlier, since these discriminatory situations and mistreatment take place within a private space (the home), they lack visibility, leading to social and political indifference towards their human and labor rights, thus contributing to a situation of systematic discrimination (CONAPRED 2011). This includes the institutional order, as there are no public policies and effective mechanisms in place to enable the full exercise of their rights (Bautista 2012).

This review of the discriminatory conditions faced by indigenous women working as domestic workers does not seek to victimize them, as efforts have been made to make them aware of their rights and to advocate for institutional changes to improve their working conditions. However, it is essential to recognize, as proposed by Durin (2017), that maintaining precarious conditions for domestic workers "is evidence of the deeply unequal nature of Mexican society, and of the patriarchal and racist regime in force" (p. 19).

4.4 Public Actions to Dignify Domestic Work: Legal Framework and Cultural Change

It is essential to consider public policy actions that allow overcoming the subordinate condition of domestic workers in general, and indigenous workers in particular, as well as the normalization of their precarious labor conditions, in favor of dignified conditions, to which all workers are entitled. Domestic workers, although they have been working under discriminatory conditions for decades, have also organized themselves in the struggle to make their labor conditions visible and dignified.

The largest organization in Mexico is the CACEH, that is National Center for Professional Training and Leadership of Domestic Workers (formerly Center of Support and Training for Domestic Workers) created in 2000 to promote and defend the labor and human rights of domestic workers in Mexico. Its leader, Marcelina Bautista, has led the fight for the recognition of their labor rights in various national and international forums. Domestic workers' associations, along with civil society organizations, have led a significant struggle to promote decent work and improve working conditions for domestic workers through institutional and legal changes.

In institutional terms, the Mexican State ratified the ILO Convention 189 in 2020, which aims to ensure that those who perform domestic work can enjoy decent employment conditions and equality with other workers (ILO 2011). It establishes that member states must adopt measures to ensure the promotion and protection of the human rights of domestic workers. This ratification supports initiatives such as the incorporation of domestic workers into the Social Security system.

Since then, some achievements and progress have been made in recognizing the labor rights of domestic workers in Mexico. In 2020, domestic work was included in the Table of General and Professional Minimum Wages, to ensure a minimum level of income and provide a benchmark for fair remuneration (Secretaría de Gobernación 2020). On July 3, 2021, ILO Convention 189 came into effect, recognizing that the protection of the rights of this sector also implies eliminating discrimination in employment and occupation, as well as providing effective protection against forms of workplace violence and gender-based violence (Secretaría del Trabajo y Previsión Social 2021).

In 2022, changes were made to the Federal Labor Law regarding social security, establishing the obligation of employers to register their domestic workers with the Mexican Social Security Institute (IMSS) based on the number of hours and days the person works. This also led to a reform of the IMSS law to ensure access for domestic workers to the insurance benefits provided by that law: (1) health and maternity insurance, (2) work-related risks, (3) disability and life insurance, 4) retirement, advanced age, and old age insurance, and (5) daycare and social benefits (IMSS n.d.). Although these changes are significant, there are currently no effective mechanisms

4.4 Public Actions to Dignify Domestic Work: Legal Framework ...

to monitor and guarantee widespread social security registration, resulting in a still very low affiliation rate.[2]

Thus, significant effort is needed to make these rights effective and move beyond mere publication of laws. In this regard, the dissemination of labor rights has been important. The CACEH organization has defined a national agenda that promotes decent work, the effective implementation of ILO Convention 189, guaranteeing the integrity and labor rights of domestic workers, promoting changes in the Federal Labor Law in favor of domestic workers, and promoting awareness campaigns for a cultural change in favor of the revaluation of domestic work. Also, it has carried out various informative campaigns about the labor rights of domestic workers, including a manual on cultural identity and non-discrimination of domestic workers (Centro Nacional para la Capacitación Profesional y Liderazgo de las Empleadas del Hogar A.C. n.d.), and together with other organizations, has promoted a manual of good practices for employers (Azuela n.d.), and on social networks promotes guides on the rights of domestic workers. Also, the Instituto de Liderazgo Simone de Beauvoir (2023) published a digital "Informative Guide on Complaint Handling and Legal Advice for Paid Domestic Workers."

Undoubtedly, the work of organizations and government actions are essential to pave the way for the full recognition of the labor rights of domestic workers. Along with these institutional actions, it is essential to promote a cultural transformation focused on eradicating negative stereotypes and systematic discriminatory treatment. This also requires recognizing the dignity and importance of their work (CONAPRED 2015), and in the case of indigenous workers, ensuring that their ethnic condition does not promote discriminatory attitudes, but rather recognition:

> The strange thing is when they ask our colleagues, "If you're being exploited, why don't you stop being a servant?" It's not about that; it's like saying, to avoid discrimination, why don't I stop being indigenous? If it's not shameful to be a servant, it's not shameful to be a laborer. What's shameful is that civil society, which has had the opportunity to educate itself, hasn't realized the importance of recognizing that other world, the indigenous one. (Sánchez et al. 2004, p. 317)

Dignifying labor conditions and promoting recognition of indigenous workers involves overcoming deep-seated and historical inequalities, moving away from the logic of charity and welfare towards recognizing the right to non-discrimination and dignified treatment. This is not an easy task, as Durin (2017) suggests, it implies both political and cultural challenges, as it "also entails transforming gender, class, and ethnic representations that contribute to the undervaluation of the occupation and the people who carry it out" (p. 378).

---

[2] It was reported that in 2023 there were in Mexico 2.5 million of paid domestic workers from 15 years of age and older in Mexico (INEGI 2024), but in September 2024 there were only 59,456 registered in the Social Security System (IMSS 2024).

4.5 Conclusion

This chapter closes by emphasizing the need to bring to light the conditions faced by domestic workers as a necessary first step towards dignifying their labor conditions and ensuring the right to non-discrimination. In the case of indigenous women, this entails recognizing their cultural particularities and overcoming racist stereotypes that foster and perpetuate conditions of inequality and discrimination. The legal reforms on labor matters represent progress in the recognition and regulation of working conditions for this sector and are the result of the struggle and effort of various social actors, particularly the actions of organized domestic workers.

The current challenge is to effectively and generally enforce these rights through institutional mechanisms, such as achieving full access to social security. This also requires a significant cultural shift toward valuing domestic work as an indispensable activity for the maintenance of life and the social reproduction of households. Similarly, efforts must be made to eradicate the stigma that associates domestic work with servitude, as well as the practices of violence, exploitation, and discrimination that afflict domestic workers in general and indigenous domestic workers in particular, based on the prejudices associated to ethnic origin.

It is a daunting task that includes institutional and cultural changes, wherein, given its relational nature, it is necessary to sensitize and raise awareness not only among employers but also to permeate different population groups in favor of effective recognition and full access to dignified treatment.

References

Arizpe, L. (1978). *Migración, etnicismo y cambio económico (un estudio sobre migrantes campesinos a la Ciudad de México).* El Colegio de México.
Azuela, M. (n.d.). *Manual de buenas prácticas para empleadoras y empleadores justos.* Hogar Justo Hogar. https://caceh.org.mx/wp-content/uploads/2023/12/Manual-de-buenas-practicas-para-empeladores-y-empleadoras-justos.pdf
Bautista, M. (2012). Hacer algo distinto para erradicar la discriminación y explotación hacia las trabajadoras del hogar. En Consejo Nacional para Prevenir la Discriminación (Comp.), *Dos mundos bajo el mismo techo. Trabajo del hogar y no discriminación* (pp. 29–38). CONAPRED.
Bautista, M. (2022). *Trabajo del hogar: invisible pero necesario.* CACEH. https://caceh.org.mx/wp-content/uploads/2018/12/Identidad-cultural.pdf
Centro Nacional para la Capacitación Profesional y Liderazgo de las Empleadas del Hogar A.C. (n.d.). *Identidad cultural y no discriminación de las empleadas del hogar.* CACEH. https://caceh.org.mx/wp-content/uploads/2018/12/Identidad-cultural.pdf
Consejo Nacional para Prevenir la Discriminación. (2008). El trato social hacia las mujeres indígenas que ejercen trabajo doméstico en zonas urbanas. *Documento de trabajo No. E-08-2008.* CONAPRED. http://cedoc.inmujeres.gob.mx/lgamvlv/CONAPRED/conapred07.pdf
Consejo Nacional para Prevenir la Discriminación. (2011). *30 de Marzo Día de las trabajadoras del hogar.* CONAPRED.
Consejo Nacional para Prevenir la Discriminación. (2015). *30 de marzo. Día internacional de las trabajadoras del hogar.* CONAPRED.

References

Consejo para Prevenir y Eliminar la Discriminación de la Ciudad de México, Centro Nacional para la Capacitación Profesional y Liderazgo de las Empleadas del Hogar A.C, Organización Internacional del Trabajo. (2021). *Informe sobre la situación de los derechos de las personas trabajadoras del hogar en la Ciudad de México*. COPRED, CACEH, OIT, Gobierno de la Ciudad de México. https://caceh.org.mx/wp-content/uploads/2022/02/informe-sobre-la-situacion-de-los-derechos-de-las-personas-trabajadoras-del-hogar-en-la-ciudad-de-mexico.pdf

Durin, S. (2017). *Yo trabajo en casa: trabajo del hogar de planta, género y etnicidad en Monterrey*. CIESAS – Publicaciones de la Casa Chata

Goldsmith, M. (1981). Trabajo doméstico asalariado y desarrollo capitalista. *Ideas feministas de Nuestra América*: https://ideasfem.wordpress.com/textos/i/i17/

González, L. (2012). Trabajo del hogar y desigualdad de género. En Consejo Nacional para Prevenir la Discriminación (Comp.), *Dos mundos bajo el mismo techo. Trabajo del hogar y no discriminación* (pp. 91–99). CONAPRED.

González, L. (2015). *Organización espacial y social de la cocina mazahua en San Antonio Pueblo Nuevo, San José del Rincón (1950–2013)* [Tesis de Maestría, El Colegio de Michoacán A.C]. Repositorio de El Colegio de Michoacán.

Gracia, M. A y Horbath, J. E. (2019). Condiciones de vida y discriminación a indígenas en Mérida, Yucatán, México. *Estudios Sociológicos, 37*(110), 277–307. https://estudiossociologicos.colmex.mx/index.php/es/article/view/1666/1786

Gramsci, A. (1975). *Cuadernos de la cárcel* (tomo 1). Ediciones Era

Gutiérrez, L. (2012). Mujeres indígenas trabajadoras del hogar. *Revista de derechos humanos - dfensor*, 1, 19–23 https://www.corteidh.or.cr/tablas/r27855.pdf

Gutiérrez, L. (n.d.). Trabajadoras del hogar indígenas en la Ciudad de México. *Las trece semillas zapatistas. Conversaciones desde los pueblos originarios*. ttps://tzamtrecesemillas.org/sitio/trabajadoras-del-hogar-indigenas-en-la-ciudad-de-mexico/

Gutiérrez, N. (2015). ¿Es una ventaja ser indígena en México en el siglo XXI? En *Ser indígena en México. Raíces y derechos. Encuesta Nacional de Indígenas* (pp. 29–164). Instituto de Investigaciones Jurídicas, UNAM.

Hernández, I. (2025). Movilidades laborales internas y metropolitanas desde comunidades rurales de la región noroeste del Estado de México. En A. Jardón (Coord.), *Escenarios de las movilidades y migraciones contemporáneas en el Estado de México*. Universidad Autónoma del Estado de México. http://ri.uaemex.mx/handle/20.500.11799/141973

Hernández, I y Jardón, A.E. (2018). Dinámicas contemporáneas de las movilidades rurales hacia las zonas metropolitanas de Toluca y Valle de México. El caso de la región noreste del Estado de México. En N. Baca, Z. Ronzón, R. Romo, R.P. Román y M. Padrón (Coords.), *Migración y movilidades en el centro de México* (pp. 171–189). CONAPO, UAEM.

Instituto Nacional de Estadística y Geografía. (2024, 26 de marzo). *Estadísticas a propósito del Dia Internacional de las Trabajadoras del Hogar* [Comunicado de prensa núm. 204/24]. https://www.inegi.org.mx/contenidos/saladeprensa/aproposito/2024/EAP_tdom.pdf

Instituto Mexicano del Seguro Social. (n.d.). *Beneficios*. IMSS. https://www.imss.gob.mx/personas-trabajadoras-hogar/beneficios

Instituto Mexicano del Seguro Social. (2024, 07 de octubre). *Puestos de trabajo afiliados al Instituto Mexicano del Seguro Social* [Comunicado de prensa]. https://www.imss.gob.mx/prensa/archivo/202410/009

Instituto de Liderazgo Simone de Beauvoir. (2023). *Cartilla informativa sobre atención de quejas y asesoría legal para personas trabajadoras del hogar remuneradas*. ILSB. https://www.gob.mx/cms/uploads/attachment/file/878020/Cartilla_ILSB.pdf

International Labour Organization. (2011). *Domestic Work Convention, 2011,* (189). ILO. https://normlex.ilo.org/dyn/nrmlx_en/f?p=NORMLEXPUB:12100:0::NO::p12100_ILO_CODE:C189

López, O. P. (2017). *Empoderamiento de las mujeres mazahuas del Estado de México. El caso de las que se quedan y las que se van de San Pedro del Rosal, Atlacomulco, 1950–1960* [Tesis de Licenciatura, Universidad Autónoma del Estado de México]. http://hdl.handle.net/20.500.11799/67656

Oehmichen, C. (2005). *Identidad, género y relaciones interétnicas. Mazahuas en la Ciudad de México*. UNAM-Instituto de Investigaciones Antropológicas.

Ojarasca. (2005). Cancionero de ausencias. *Ojarasca. La Jornada*, 95. https://www.jornada.com.mx/2005/03/21/oja95-mazahua.html

Ramírez, T. (2025). Cambios y continuidades en los patrones migratorios y movilidades poblacionales en el Estado de México. En A. Jardón (Coord.), *Escenarios de las movilidades y migraciones contemporáneas en el Estado de México*. Universidad Autónoma del Estado de México. http://ri.uaemex.mx/handle/20.500.11799/141973

Rodríguez, J. (2023). *Una teoría de la discriminación*. Universidad Autónoma Metropolitana-Iztapalapa.

Sánchez, C. (2004). La diversidad cultural en la Ciudad de México. Autonomía de los pueblos originarios y los migrantes. En P. Yanes, V. Molina y O. González (Coords.), *Ciudad, pueblos indígenas y etnicidad* (pp. 57–87). Universidad de la Ciudad de México.

Sánchez, P., González, M., Ayala, B., Gutiérrez, L., De la Torre, K., Ventura, B. y Hernández, F. (2004). Sobre la experiencia y el trabajo de las organizaciones indígenas en la ciudad de México. En P. Yanes, V. Molina y O. González (Coords.), *Ciudad, pueblos indígenas y etnicidad* (pp. 321–368). Universidad de la Ciudad de México.

Secretaría de Gobernación. (2020, 17 de diciembre). *Un logro, la inclusión de personas trabajadoras del hogar y agrícolas en la lista de salarios mínimos: Conapred*. [Comunicado de prensa]. https://www.gob.mx/segob/prensa/un-logro-la-inclusion-de-personas-trabajadoras-del-hogar-y-agricolas-en-la-lista-de-salarios-minimos-conapred

Secretaría del Trabajo y Previsión Social. (2021, 03 de julio). *Entrada en vigor del Convenio 189 de la Organización Internacional del Trabajo protege a las personas trabajadoras del hogar* [Comunicado de prensa]. https://www.gob.mx/stps/prensa/comunicado-conjunto-017-2021?idiom=es

Chapter 5
Labor Discrimination: The Case of Young People Who Study and Work

In the SDGs and the MC, special attention is paid to the young population, aiming for them to enjoy peace and prosperity, ensuring their security, integrity, full exercise of human rights, availability of options, access to health, education, and social protection, among others (CEPAL 2013; NU n.d.b). Despite these international efforts, the prevalence of discrimination among young people continues. In the case of Mexico, 22.5% of young people aged between 18 and 24 reported experiencing some form of discrimination in 2022, with the three main areas being the street or public transportation (41.23%), work or school (35.1%), and social networks (22.9%) (INEGI 2022). These data highlight the need to analyze discrimination among the young population, not only to identify and describe the factors that contribute to this social problem but also to make visible its impact.

Therefore, this chapter proposes to present the case of young people in the State of Mexico who are studying at the undergraduate level and working, as a way to approach discrimination in the workplace. The importance of selecting this case lies in the fact that studying and working simultaneously is not always done under ideal conditions. Young people pursuing their professional education face various obstacles that limit their ability to obtain decent work and exercise their rights freely. To a large extent, these restrictions come from the labor market, which makes labor discrimination evident.

The development of this chapter is divided into four sections. The first section presents studying and working as fundamental rights in the lives of young people, both recognized in the international agenda such as the SDGs and the MC. The second section addresses positions and components that arise from studying and working simultaneously, highlighting that young people develop this strategy, but a broader context influences it. The third section illustrates the harmful mechanisms of the labor market that restrict access and inclusion of young people who study and work simultaneously, generating discriminatory actions for the benefit and under the protection of the same market. Finally, the last section presents some experiences

© The Author(s), under exclusive license to Springer Nature Switzerland AG 2025
A. E. Jardón Hernández et al., *Multiple Discriminations*,
SpringerBriefs in Environment, Security, Development and Peace,
https://doi.org/10.1007/978-3-031-85826-0_5

of undergraduate students at the Autonomous University of the State of Mexico (UAEMéx), highlighting the dilemmas and challenges they face when studying and working, especially when they encounter limitations that prevent them from fully engaging in employment and their labor rights are violated due to simultaneous engagement in both activities.

5.1 Study and Work: Fundamental Rights in the International Agenda

Youth is a psychosocial stage where the main transitions or changes occur, defining individuals' future trajectories such as leaving home to become independent, starting their own families, completing studies, professionalizing, transitioning from education to work, entering the labor market, among others (Chacaltana et al. 2018). Additionally, this age group is considered a social, economic, political, and cultural force for the present and future of the country (Observatorio de la Juventud Iberoamericana [OJI] 2019). However, it is worth observing the contexts in which young people live and the conditions they have to develop their potential. Here, two relevant events for this stage of life are highlighted: education and work.

In Goal 4 "Quality Education" of the SDGs, the importance of education is expressed, indicating that it is the key to achieving other SDGs, as it can break the cycle of poverty, help reduce social inequalities, achieve gender equality, lead to a healthier and more sustainable life, promote tolerance among people, and contribute to the development of more peaceful societies (NU n.d.). Therefore, education "is a human right for all throughout life" (Organización de las Naciones Unidas para la Educación, la Ciencia y la Cultura [UNESCO] n.d.). Additionally, formal education can contribute to upward social mobility in individuals, considering factors that influence it, such as social background (Solís and Dalle 2019). Due to the relevance of education in the lives of young people, actions must be taken towards quality, equitable education that ensures equal access.

In Goal 8 "Decent Work and Economic Growth," decent work is defined as work that is productive and provides fair income, workplace security, and social protection for families, as well as better prospects for personal development and social integration (NU n.d.). Thus, work is considered the means to earn income, achieve goals, assume responsibilities, and make future commitments; it is "the path that enables social and economic inclusion" (OJI 2019, p. 39). Due to its importance, Goal 8 includes two universally valid targets and one specific target associating the young population with work (NU n.d.),

- 8.5 By 2030, achieve full and productive employment and decent work for all women and men, including young people and persons with disabilities, as well as equal pay for work of equal value.
- 8.6 By 2020, substantially reduce the proportion of youth not in employment, education, or training.

5.1 Study and Work: Fundamental Rights in the International Agenda

- 8.b By 2020, develop and implement a global strategy for youth employment and apply the Global Jobs Pact of the ILO.

In all three goals, the need to promote equal opportunities and treatment for young people is evident, while acknowledging that this population group has age-specific characteristics. It is also recognized that the historical and contextual conditions in which this population group lives have led to inequalities in labor market inclusion, which are sought to be addressed through these international parameters.

On the other hand, in the Montevideo Consensus on Population and Development (CEPAL 2013), priority measure B "Rights, needs, responsibilities, and demands of children, adolescents, and youth" is proposed. It recognizes that these population groups are rights holders and development actors, aiming to ensure opportunities for their full development free from discrimination. Regarding work and education, the agreements emphasizing these rights are:

- 9. Investing in youth through specific public policies and differential access conditions, especially in public, universal, secular, intercultural, discrimination-free, free, and quality education, to ensure that it becomes a stage of full and satisfying life. This enables them to build themselves as autonomous, responsible, and solidarity-oriented individuals, capable of creatively confronting the challenges of the twenty-first century.
- 10. Promoting and investing in labor and employment policies and special training programs for youth that enhance collective and personal capacity and initiative, and enable the reconciliation of studies with work activities, without precarization of work and ensuring equality of opportunities and treatment (CEPAL 2013, pp. 15–16).

As can be noted, both education and work are presented as means to achieve favorable conditions for young people, such as a full and satisfying life. Likewise, in agreement number 10, emphasis is placed on promoting public policies that harmonize both activities without implying harmful effects on their conditions, as the goal is to ensure the free exercise of their rights under any circumstances and not a violation of these rights.

The international recognition of young people as rights holders commits various entities to design, implement, and evaluate public actions aimed at promoting equal opportunities and treatment in the exercise of their human rights. However, it is important to consider that this stage of life entails differentiated conditions compared to other population groups, which should not be understood as obstacles that limit or deny rights; on the contrary, they should be taken into account for the benefit and enjoyment of individuals. Thus, studying and working are fundamental rights for the full development of young people. To continue the discussion, the following positions and components are presented, distinguishing them when undertaking simultaneous study and work actions, to envision the tensions in the lives of young people within a broader contextual framework beyond the individual.

5.2 Challenges and Dilemmas of Studying and Working in the Young Population

Challenges and dilemmas of studying and working with the young population are acknowledged due to their complexity and the diverse perspectives surrounding them. From a positive perspective, the combination of study/work, and theory/practice relationships allows for linking learning experiences and enhancing competencies (Bustamante et al. 2018). However, from a negative standpoint, it is indicated that studying and working is one of the main risk factors for school dropout (Cruz et al. 2017). Instead of being an asset to increase capacity development and opportunities, it becomes a limitation.

Both positions result from the wide diversity of environments, resources, decisions, priorities, interests, and more, of young people engaging in these activities. To delve into the topic, three components are presented here that converge in the heterogeneity and complexity of scenarios. Firstly, Cruz et al. (2017) indicate that it is not enough to simply need to study and work; a structure of opportunities is required for it to happen, which depends on the market, the state, and society. The labor market can have an impact on access to work and insertion expectations, such as the existence of jobs in the service sector, the informal sector, or family businesses, which may attract young people who are studying. Regarding the state, the authors highlight its direct impact, as the absence of educational infrastructure in residential areas or low-quality education can inhibit youth school insertion and lead to the development of labor strategies. Finally, there is society, which is "represented by the community and the family as agents of support for socialization, social integration, and identity formation" (Cruz et al. 2017, p. 577). However, according to the authors, society can also limit access to education or work and reproduce social inequities, such as truncating the educational trajectory of male children to join the labor market and generate income for the household.

With the contribution of Cruz et al. (2017), it can be identified that studying and working not only depends on individual factors or household strategies but also the broader context. That is, the conditions under which a person simultaneously engages in study and work activities are influenced by the political, economic, and social convergences at different levels, such as individual, family, local, state, regional, national, and even global levels.

In the second component, three practices stand out when engaging in simultaneous study and work activities. Bustamante et al. (2018) point out, on one hand, the way activities are prioritized, on the other hand, the distribution of time and effort, and finally, the decision-making process to opt for more work or more study hours. From this perspective, young people face dilemmas when engaging in both activities, which implies constant decision-making regarding the challenges that can be caused by factors, both internal and external, such as the preference for work, the work environment, varying school workload, fatigue, the distance between places of residence, school, and work, among others.

Finally, the third component is the re-signification given to each activity. Based on a study of young people studying and working in a supermarket in Argentina, Guiamet and Saccone detected modifications in the representations of each activity. "If initially work appeared as a means to study, over time it also becomes a means for other changes that occur in their lives" (2013, p. 241). Thus, moving away from the original home, forming a partnership, having a child, job promotion, and school scholarships, among others, can lead to a change in the importance of one activity over another, both in interests and in dedication and distribution of time.

From the three components mentioned, two central reflections emerge. First, studying and working is a matter of social interest and intervention rather than individual, as decisions, priorities, and tensions respond to the resources and factors available in the immediate environment and broader contexts such as economic, political, and social factors. Second, studying and working is a dynamic process, not only due to the entry and exit in each of the activities but also due to the meaning assigned to them and the constant reassessment and organization of time, resources, priorities, efforts, and more, involved in the concurrency of events.

In this way, it is understood that studying and working is a strategy implemented for personal purposes but influenced by broader contexts. Therefore, it is not far-fetched to think that the same environment fosters the defense or violation of human rights for young people. Hence, the structure supporting the development of both activities must be based on the recognition of these rights and a multi-level regulatory framework that promotes their protection and enjoyment. Taking this scenario into consideration, mechanisms harmful to the labor market that undermine the full inclusion and attainment of decent work for the population sector under study are presented below.

5.3 Studying and Working: A Strategy Limited by the Labor Market

According to the Population and Housing Census of 2020, there were a total of 14,736,111 individuals aged between 18 and 24 years nationwide (INEGI 2020a). This group represented 14.7% of the total population. Two decades ago, the proportion was 18.7% (INEGI 2000), indicating a shift in the country's demographic structure towards an aging population, although the young population still holds significant weight. Within this group, 743,719 individuals reported attending university to pursue a bachelor's degree while being economically active (INEGI 2020a). In other words, 5% of the country's young population declared studying and working simultaneously. This data is noteworthy because 15.3% of the same age group were solely pursuing a bachelor's degree, and 46.8% were exclusively employed.

For the State of Mexico, the most populous entity in the country with 13,742,847 inhabitants, the group of young people aged 18–24 consisted of

2,004,632 individuals, of which 49.9% were women and 50.1% were men (INEGI 2020a). The proportion of young people studying for a bachelor's degree while being economically active simultaneously does not differ from the national context, at 4.8% (96,077 young people), with 47% being women and 53% men (INEGI 2020a).

The percentages of young people exclusively studying or working demonstrate their insertion in both areas, albeit with their respective constraints. However, the percentages of young people engaged in both activities nationally and statewide hover around 5%, indicating a clear limitation. With the available data, two problems become apparent. On one hand, there may be an underreporting of labor activities, such as not considering jobs outside the formal sector, temporary employment, positions with variable compensation, or simply due to the respondents' lack of knowledge within the household. However, given the low percentages, another type of problem may be underlying, one that has a structural nature within the labor market.

In Latin America and the Caribbean region, including Mexico, young people continue to face obstacles to their full inclusion in the labor market (Chacaltana et al. 2018). Indeed, the young population is considered among the most vulnerable in this regard (Weller 2007; Vargas and Cruz 2014; Urrutia and Cuenca 2018). Their insertion and inclusion are marked by notable precarization of labor conditions, which could explain why studying and working is not a viable strategy for young people, even if they need it, as mentioned by Cruz et al. (2017). Contextual and structural factors, rather than personal ones, limit opportunities to secure employment while engaging in another relevant and demanding activity such as pursuing a professional career.

Institutions involved in this strategy do not create a structure of opportunities (Cruz et al. 2017); on the contrary, they impose limits and exclusions on the free exercise of rights. Specifically, the labor market produces mechanisms of exclusion or limited insertion, such as legally allowing remuneration below the law, offering temporary jobs without benefits, implementing temporary contracts, and limiting social security guarantees, among other actions that lead to labor discrimination.

It is argued here that the labor market itself perpetuates discrimination, as its structure and regulatory framework endorse mechanisms of exclusion and differential treatment. In Convention 111, presented in the conceptual chapter of this book, the right to equal opportunity and treatment is mentioned, with only three exceptions (OIT 2014). However, these exceptions blur the line between what is considered a requirement for job performance versus a discriminatory motive. Job postings are a clear example of this, as they may demand attenuating requirements specific to a certain phenotypic model, without these being necessary for performing the job's functions (Rodríguez 2023).

With job postings, one can also observe the complex relationship between the offering and demand sides of the labor market, where legal and institutional support leans towards the employer sector. In the first place, Soberanes (2022) indicates that job postings do not constitute employment discrimination because there is no employment relationship; it is only a public offer. However, discriminatory acts may

5.3 Studying and Working: A Strategy Limited by the Labor Market

be established by preventing someone from getting a job due to reasons contrary to their dignity. But, at the same time, the author points out that job postings involve the prohibition of treating a job candidate unequally and the employer's freedom to enter into a contract with whomever they wish, without an obligation of compliance (Soberanes 2022).

The above demonstrates the paradoxes created by the imprecisions between what constitutes a discriminatory manifestation of a valid and legally regulated resource. This results in the exclusion of young people from the labor market who do not meet the requirements stipulated by employers, which are unnecessary for performing a job, without the possibility of legitimate defense of their rights and equal opportunities.

Similarly, the issue of contracts presents itself. According to García (2010), job instability is expressed through the scarcity of written contracts and an increase in temporary contracts, constituting one of the most vulnerable aspects for salaried workers. This situation has significant effects on career paths, as it does not generate seniority in social security, a constant income flow, or the possibility of promotion to higher positions. From the author's research results, it is demonstrated that young people with lower levels of education are the ones who lack written contracts the most; this is related to lower income and benefits compared to those with permanent written contracts (García 2010).

Furthermore, it is worth noting that, following labor reforms in the 1980s towards flexibilization, new forms of employment contracts were introduced: trial periods, initial training contracts, seasonal contracts, part-time contracts, and subcontracting. These "are forms of employment that transfer risks to the worker and cause uncertainty, as they do not guarantee job security" (Quiñonez and Rodríguez 2015, p. 196). In general, the lack of a permanent written contract implies the violation of labor rights and the precariousness of working conditions for young people who are just starting their work trajectory.

With job postings and contracts, it can be observed that the labor market system has generated a series of instruments under the protection of the regulatory framework that, directly or indirectly, endorse the violation of people's labor rights. The needs of the market are prioritized in pursuit of global competitiveness, at the expense of fundamental rights. Consequently, "young people currently face greater difficulty in obtaining and maintaining a protected job, due to the limited options for jobs that provide decent pay and social protection" (Vargas and Cruz 2014, p. 216). This situation worsens when young people simultaneously engage in another activity such as studying a professional career, as the system imposes further restrictions on their full inclusion.

Therefore, it is argued here that the decisions to study and work simultaneously by young people, as well as the circumstances in which both activities are carried out, are directly influenced by structural factors. Thus, to provide evidence supporting this position, the following section presents the results of a research project that captured the experiences of young people studying for a university degree and working in the State of Mexico.

5.4 Experiences of Young People Who Study and Work at the Autonomous University of the State of Mexico

On March 23, 2020, Mexican authorities implemented the National Healthy Distance Campaign, which involved basic prevention measures, temporary suspension of non-essential activities, limitation of mass gathering events, and special attention to older people, aimed at preventing the spread of COVID-19 (Gobierno de México 2020). In this scenario, a project was conducted to identify the contexts in which young people from UAEMéx study and work, as well as their various understandings of work. However, it was inevitable to also address, in the interviews, the impacts caused by the sanitary measures on the labor trajectories of the target population, as well as their perceptions and experiences regarding the labor market. The findings revealed the complications faced by young people, not only in an unprecedented historical moment but also in broader terms concerning their relationship with the labor market, characterized by violations of labor rights.

Within the framework of the project, 25 semi-structured interviews were conducted with young people between 18 and 24 years old who were enrolled in a professional career at UAEMéx and were working during the observation period. The results distinguished three significant events for young people: university enrollment, work, and the pandemic. Based on these events, three types of trajectories were identified: A, B, and C. In trajectory A, young people entered college, then started working, and then the COVID-19 pandemic emerged. In trajectory B, they first started working, then entered college, and then the pandemic emerged. In typology C, they first entered college, then the pandemic emerged, and later they started working (Fig. 5.1).

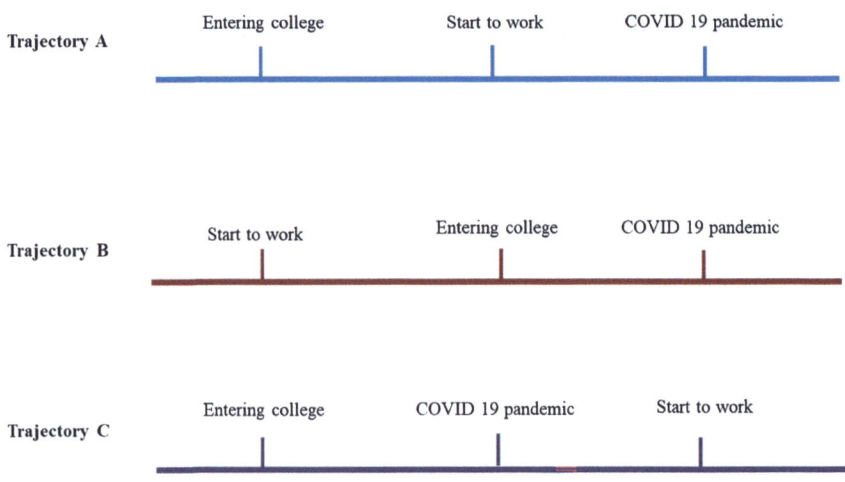

Fig. 5.1 Pathways of young people who study and work. *Source* Own elaboration based on 25 interviews conducted with young people who study at UAEMéx and work, 2020–2021

5.4 Experiences of Young People Who Study and Work at the Autonomous ...

As can be seen, each trajectory entails particular conditions and experiences among the interviewed young people. To avoid presenting each case individually, here are some predominant elements. In trajectory A, young people prioritize university studies, and work tends to be a secondary activity only to support expenses associated with studying. Those in this trajectory share the following characteristics:

- They enroll in university immediately after completing high school.
- There is not much clarity in the choice of the selected major.
- They start working to cover university expenses.
- Jobs are part-time, on weekends, hourly, or project-based.
- The employment relationship is not formal (no written contract or social security).
- Incomes are variable.
- They consider work as a way to gain experience, even if it is not very profitable.
- Study is the priority, both in attention and in hours dedicated per week.

Teresita's experience is an example of trajectory A. She began working in the second semester of her Bachelor's degree in Communication. Her job was as a waitress at special events, working 6–10 h on weekends. She mentions that her income varied between 500 and 1000 pesos ($25 and $50 dollars approximately) per event, and depending on extra compensations (tips). Additionally, she was not called to work every weekend, and sometimes she covered two events on the same day. There was no written contract or benefits. This case may illustrate the existence of jobs compatible with class schedules, providing income to cover her studies, as Teresita mentioned. The work only serves the purpose of providing monetary resources but does not offer other specialized skills compatible with her professional training.

In Teresita's case, working as a waitress at special events is a way to make generating monetary resources compatible with her undergraduate studies. However, the occasional nature and variability of her income, combined with the precariousness of working conditions, pose a constant risk of losing her job, leading to a violation of the right to opportunity and fair treatment as stipulated in Convention 111 (OIT 2014). Whether hourly or for temporary activities, access to the enjoyment of labor rights is available.

In trajectory B, there is a greater emphasis, both in terms of time and interest, on work rather than university studies. Here, it is worth revisiting the distinction made by Garabito (2018) between university student workers and university students who work, where "the distinction focuses on how the timing of insertion into one or the other sphere (school and work) influences the organization of daily life" (p. 103). Trajectory B includes university student workers and is characterized by:

- They enroll in university after a year or more after finishing high school.
- They have a clearer choice of major.
- They start university to obtain a degree.
- Jobs are part-time and full-time.
- Employment is in family businesses or established companies.

- The employment relationship is more formal (hospitals, city councils, etc.)
- Incomes are fixed and considered adequate.
- Work is the priority, both in attention and in hours dedicated per week.

The case of Karen exemplifies trajectory B. She studied a technical career in nursing and started working at a city hospital. Her entry into the hospital was through a public call and she obtained an indefinite position. She mentioned that her job was in the night shift and she had to cover 36 hours per week. The income was enough to cover her expenses, and she had all the legal benefits. However, she indicated that although studying and working was challenging, she managed her time to do both activities. If she encountered difficulties, she would ask the University for permission to miss classes or, if necessary, to take a leave of absence for a while, although she had not used these resources during the six semesters she had completed. This case allows us to understand that a Bachelor's Degree in Social Work is a second profession that complements nursing. However, the work in the hospital does not depend on the accreditation (degree) of the current career. Therefore, it can be indicated that studying with a stable job is a possibility for diversification and personal development.

Similarly to Karen's case, there are other examples where work takes priority over study, but the reasons are diverse. Mario started working at the age of 13 in the family business of making and selling barbecue. The work was on weekends and involved 16–20 h in total. His income was less than 500 pesos ($25 dollars approximately) per week, but his parents supported him with materials for university. Although there was no overlap in schedules between study and work, in the interview he mentioned that the priority was the family business, as several families depended on it, and he saw himself in the future managing the business like his parents did. Here it can be seen that work is not due to the need for remuneration to cover university costs, but as support and continuity for the family business, yet university studies are integrated into daily dynamics to pursue a profession.

Those who find themselves in the conditions described in trajectory B increase their chances of experiencing mismatches between studying and working, such as not knowing how much study time the career will require, which implies making modifications once enrolled so that studying requires less time (Bustamante et al. 2018). This shows that the labor market system influences young people to prioritize work over study. Thus, the labor market restricts the right to training and professional development, which is another expression of labor discrimination.

Finally, in trajectory C, young people started their university studies, and then the pandemic emerged, which led to entry into the labor market. This evidences an action plan in response to the crisis caused by the COVID-19 pandemic, either due to the loss of the main source of household income, the death of a family member, or as a response to confinement. According to Montiel et al. (2021), the COVID-19 crisis has been catastrophic in many ways, but it is also a driver to seek alternatives and solutions to reduce its impact. The incorporation of university students into the labor market implied a personal and household strategy, but not always under the best conditions, especially in a moment of global constraint. Generally, the characteristics of young people in this trajectory are:

- They enter university immediately after completing high school.
- There isn't much clarity in the choice of the selected career.
- They did not work before the pandemic.
- They started working based on household conditions during the pandemic.
- Jobs are part-time or full-time.
- Incomes are variable.
- Incomes are entirely destined to cover household expenses.
- Work becomes the priority, both in attention and in hours dedicated per week.

Carolina's experience exemplifies her entry into the workforce during the COVID-19 pandemic. She was in the second semester of her Communication degree when the health crisis was declared, leading to associated measures. Her father's layoff prompted her to seek work as an assistant in a beauty salon. She worked from Monday to Friday, but her hours and income varied depending on demand and tasks. She did not have a written contract or benefits, only monetary compensation, which helped cover some university expenses, especially internet service for her mobile phone and the computer she shared with her sister.

Carolina's experience highlights the fragility of labor regulations during moments of crisis, although this is not new. Labor flexibilization promoted in Mexico since the 1980s has led to changes in the labor market, aiming to make quick adjustments and reduce costs for companies, but at the expense of subordinated and remunerated workers (Trejo 2017). Thus, the COVID-19 pandemic and lockdown measures only exacerbated existing poor working conditions that have persisted for decades. The violation of labor rights became more evident under the pretext of employment constraints in the country, manifested by temporary or permanent closures of jobs without the right to salary payment, compensation, or severance pay; also, work hours, salaries, and benefits were reduced (INEGI 2020b).

Through the interviews and the characteristics of the three outlined trajectories, the personal tensions caused by studying a university degree and working are visible. Young people make decisions about time dedication, priorities, and strategies to carry out both activities. However, some conditions do not solely reside at the personal level and have been considered as such. For example, a young person working 20 h a week, devoting the rest of the time to school activities, does not imply renouncing their labor rights due to time constraints. On the contrary, they are entitled to rights, regardless of time, activity, labor condition, or specific contexts such as the COVID-19 pandemic.

Among the 25 interviewees, only four indicated they had all legal benefits, and only two of them signed a written contract. These data denote the precarious conditions of the observed group, which are not due to personal decisions but to the restrictions and limitations generated by the labor market to the detriment of the labor rights of young people. In other words, studying and working or engaging in other simultaneous activities such as recreational, artistic, or caregiving should not be an excuse for carrying out discriminatory actions.

The denial of a right, as indicated, is a discriminatory act (NU 2001). In the cases addressed, young people have been denied the right to have labor rights because they lack time availability, years of experience, a title or document validating specialized knowledge, or meeting the requirements of calls, even if they are unnecessary for the job functions, such as age and gender (Nuvaez 2019). This group of individuals faces a complex system of violation of their fundamental rights, both directly and indirectly or structurally (Rodríguez 2023), as well as multiple or intersectional (Salomé 2017), which undoubtedly causes personal and social harm.

5.5 Conclusion

The conditions in Mexico for young people studying for a university degree and working are unfavorable. On the contrary, the complex structure of the labor market contributes to exclusion and limits their full inclusion, affecting their personal and professional development. These conditions restrict their rights to decent work and quality education, which are central goals in international frameworks such as the SDGs and the Montevideo Consensus.

It is argued here that the Mexican labor market produces exclusion and precarity for young people who study and work simultaneously, as the lack of access to jobs that fit their available time, along with labor and social protection deficiencies, insufficient wages to achieve their goals, lack of written contracts, and temporary jobs, among others, hinder their full integration into the labor market and violate their fundamental rights.

Studying and working simultaneously produces challenges for young people, resulting in personal tensions, complex decision-making regarding priorities, and the redefinition of both activities over time and by their social and family environment. Therefore, it is necessary to create the necessary infrastructure of opportunities to support young people and improve their participation, not only in education and work but also in political, economic, and social spheres (Chacaltana et al. 2018).

Therefore, it is important to recognize, describe, and analyze this population group in depth to identify the obstacles faced by young people who study and work. "For working students, it is a challenge to maintain academic performance and continue their educational path due to reduced study time and rigid work schedules" (Cruz et al. 2017).

In Mexico, some efforts have been made to address the needs of young people, such as the national program "Youth Building the Future." This program aims to provide training in companies and workplaces for young people between 18 and 29 years old to develop their skills and abilities. During the 12-month training, they receive financial support and medical insurance (Gobierno de México 2022). While this program can catalyze labor integration, it excludes those who study, leaving the analyzed group outside the possible benefits of the program.

In this way, the structure of the labor market, as well as its legislation, must consider the diversity of conditions in which young people are engaged, to promote the right to equal opportunities and treatment. Policies and actions must be created and implemented effectively and efficiently to fully protect those who combine significant activities such as working and studying, as well as to foster the acquisition of knowledge, link theory with practice, and promote greater employability opportunities.

Finally, the goal is to promote a culture of respect for the free exercise of rights for all individuals, particularly focusing on young people in this case. Labor regulations must align with the protection of human rights as stipulated by the OIT (2014), not undermine them. A lack of experience should not be punished with low wages or the elimination of legal benefits. The simultaneity of activities should not be an excuse to precarize work. Youth should not be a stage of life prone to vulnerability.

References

Bustamante, L., Ayllón, S. y Escanés, G. (2018). Abordando la trayectoria universitaria desde el pensamiento complejo. *Praxis Educativa*, 22(3). https://cerac.unlpam.edu.ar/index.php/praxis/article/view/2648/3217

Chacaltana, J., Dema, G. y Ruiz, C. (2018). El futuro del trabajo que queremos. La voz de los jóvenes y diferentes miradas desde América Latina y el Caribe. *Perfiles educativos*, XL(159), 194–210. https://perfileseducativos.unam.mx/iisue_pe/index.php/perfiles/article/view/58775

Comisión Económica para América Latina y el Caribe. (2013). *Consenso de Montevideo sobre población y desarrollo*. Comisión Económica para América Latina y el Caribe. https://www.cepal.org/es/publicaciones/21835-consenso-montevideo-poblacion-desarrollo

Cruz, R., Vargas, E., Hernández, A. y Rodríguez, O. (2017). Adolescentes que estudian y trabajan: factores sociodemográficos y contextuales. *Revista Mexicana de Sociología* 79(3), 571–604. https://revistamexicanadesociologia.unam.mx/index.php/rms/article/view/57679/51146

Garabito, G. (2018). Trabajo y juventudes universitarias en México: tendencias y complejidades. En A. Corica, A. Freytes y A. Miranda (Coords.), *Entre la educación y el trabajo: la construcción cotidiana de las desigualdades juveniles en América Latina* (pp. 93–108). CLACSO.

García, B. (2010). Inestabilidad laboral en México: el caso de los contratos de trabajo. *Estudios Demográficos y Urbanos,* 25(1), 73–101. https://estudiosdemograficosyurbanos.colmex.mx/index.php/edu/article/view/1368/1361

Gobierno de México. (2020). *Jornada Nacional de Sana Distancia*. Gobierno de México. https://www.gob.mx/salud/hospitalgea/documentos/jornada-nacional-de-sana-distancia

Gobierno de México (2022). Jóvenes Construyendo el Futuro. *Programas para el bienestar*. Gobierno de México. https://programasparaelbienestar.gob.mx/jovenes-construyendo-el-futuro/

Guiamet, J. y Saccone, M. (2013). Entre la educación y el trabajo: experiencias formativas de jóvenes trabajadores. *Avá. Revista de Antropología,* 22, 229–248.

Instituto Nacional de Estadística y Geografía. (2000). *Censo de población y vivienda 2000*. INEGI. https://www.inegi.org.mx/programas/ccpv/2000/#tabulados

Instituto Nacional de Estadística y Geografía. (2020a). *Censo de población y vivienda 2020*. INEGI. https://www.inegi.org.mx/programas/ccpv/2020/#tabulados

Instituto Nacional de Estadística y Geografía. (2020b). *Noticia. Encuesta Telefónica sobre COVID-19 y Mercado Laboral (ECOVID-ML) abril – julio de 2020*. INEGI. https://www.inegi.org.mx/app/saladeprensa/noticia/6207

Instituto Nacional de Estadística y Geografía. (2022). *Encuesta Nacional sobre Discriminación (ENADIS) 2022*. INEGI. https://www.inegi.org.mx/programas/enadis/2022/

Montiel, O., Flores, A., Ávila, E. y Sierra, S. (2021). "Tengo que sobrevivir": Relato de vida de tres jóvenes microemprendedores bajo COVID-19. *Telos* 23(1), 67–81. https://doi.org/10.36390/telos231.06

Naciones Unidas. (2001). *Conferencia Mundial contra el Racismo, la Discriminación Racial, la Xenofobia y las Formas Conexas de Intolerancia*. NU. https://www.un.org/es/conferences/racism/durban2001

Naciones Unidas. (n.d.). Objetivos de Desarrollo Sostenible. NU. https://www.un.org/sustainabledevelopment/es/objetivos-de-desarrollo-sostenible/

Nuvaez, J. (2019). La discriminación laboral en razón del género y la edad en Colombia. *Revista Arbitrada Interdisciplinaria Koinonía*, 4(7), 308–320. https://doi.org/10.35381/r.k.v4i7.207

Observatorio de la Juventud en Iberoamérica. (2019). *Encuesta de jóvenes en México 2019*. OJI. https://oji.fundacion-sm.org/nuestros-estudios/encuesta-mexicana-de-la-juventud/

Organización de las Naciones Unidas para la Educación, la Ciencia y la Cultura. (n.d.). *La educación transforma vidas*. UNESCO. https://www.unesco.org/es/education

Organización Internacional del Trabajo. (2014). *Guía sobre las normas internacionales del trabajo*. OIT. https://www.ilo.org/wcmsp5/groups/public/---ed_norm/---normes/documents/publication/wcms_246945.pdf

Quiñonez, C. y Rodríguez, S. (2015). La reforma laboral, la precarización del trabajo y el principio de estabilidad en el empleo. *Revista Latinoamericana de Derecho Social* 21, 179–201. https://doi.org/10.22201/iij.24487899e.2015.21.9768

Rodríguez, J. (2023). *Una teoría de la discriminación*. Universidad Autónoma Metropolitana-Iztapalapa.

Salomé, L. (2017). La discriminación y algunos de sus calificativos: directa, indirecta, por indiferenciación, interseccional (o múltiple) y estructural. *Pensamiento constitucional*, 22(22), 255–290. https://revistas.pucp.edu.pe/index.php/pensamientoconstitucional/article/view/19948/19969

Soberanes, J. (2022). La discriminación en las convocatorias laborales. *Revista latinoamericana de derecho social,* 35, 271–296. https://doi.org/10.22201/iij.24487899e.2022.35.17279

Solís, P. y Dalle, P. (2019). La pesada mochila del origen de clase. Escolaridad y movilidad intergeneracional de clase en Argentina, Chile y México. *Revista Internacional de Sociología*, 77(1), 1–17. https://revintsociologia.revistas.csic.es/index.php/revintsociologia/article/view/1018/1332

Trejo, A. (2017). Crecimiento económico e industrialización en la Agenda 2030: perspectivas para México. *Revista Problemas del Desarrollo*, 188(48), 83–111. https://www.probdes.iiec.unam.mx/index.php/pde/article/view/56026/51495

Urrutia, C. y Cuenca, R. (2018). *Las desigualdades laborales que enfrentan los jóvenes en Lima metropolitana*. Instituto de Estudios Peruanos.

Vargas, E. y Cruz, R. (2014). Búsqueda de empleo entre jóvenes de acuerdo con su participación y protección laboral en México. *Papeles de Población,* 20(81), 213–245. https://rppoblacion.uaemex.mx/article/view/8352

Weller, J. (2007). La inserción laboral de los jóvenes: características, tensiones y desafíos. *Revista de la CEPAL,* 92, 61–82. https://repositorio.cepal.org/entities/publication/70f4571b-0978-43ac-8faf-0ea31395403a

Chapter 6
Older People as Subjects of Discrimination Due to a Lack of Oversight of their Rights

In this book, various population groups that have been subjects of social analysis for over a century have been addressed, not only from academia but also from public policy, and have been present in the struggle for the recognition of Human Rights. Now, in this chapter, the aim is to observe a particular population group that only became an urgent need to address at the end of the twentieth century: the age group of 60 years and older, the ageing, who, as previously discussed in the first part, have faced different types of discrimination in their social contexts, from the family and community to the legal sphere.

Old age and aging, like other stages of life, have initially been viewed from a biological perspective, but the Social Sciences and Humanities have managed to show that there is a sociocultural construction around this stage of life and that this construction is linked to the social role assigned to older persons.

The social conditions that have led to constant discrimination are those that allow for the violation of their rights and which, today, are part of the global agenda. As demonstrated earlier, they are also present in the SDGs 2030 and regionally, have a specific section in the Montevideo Consensus, as stated in the first part of this book.

Now, the task is to show the local-regional reality in which the older population is vulnerable in three main areas: health, work, and daily life. While the data presented reflect the national panorama, the focus is on the State of Mexico.

In this chapter, there is, first, a brief overview of the trajectory of recognizing the rights of the ageing, which, although framed within the SDGs 2030 and the Montevideo Consensus in the book, here emphasizes different specific documents on aging and the elderly. Then, some basic elements to understand the phenomenon of demographic aging are presented, followed by highlighting some variables that imply discrimination in this age group. Finally, first-hand information obtained in the field during different periods of fieldwork between 2020 and 2023 is addressed through ethnographic work and open and semi-structured interviews.

6.1 Older People on the Agenda for the Recognition of Their Rights

The journey towards the recognition of this age group as rights-holders has been relatively accelerated, considering that it was in 1982 when the First World Assembly on Ageing was held in Vienna. This assembly not only aimed to raise awareness about the reality of demographic aging at the time but also emphasized the need to defend the rights of older people. It was affirmed at that time that "the situation of discrimination, mistreatment, and even abandonment that this population group sometimes suffers was not highlighted until the last decades, a recognition that allowed overcoming the conception of older people as objects of rights to consider them as rights-holders" (OPS 2023, p. 2).

Thus, in December 1990, the United Nations General Assembly designated October 1 as the International Day of Older People, and by 1991, the United Nations Principles for Older People were established (NU 1991). These principles aimed to recognize the rights of older people globally and were oriented toward five areas: independence, participation, care, self-fulfillment, and dignity. Governments were urged to integrate policies covering these five areas (NU 2010).

These principles led to the generation of the International Madrid Plan of Action on Ageing in 2002, which aimed to establish lines of action for the signatory States to protect their rights and generate conditions for their exercise. These were the main global initiatives created at the end of the twentieth century that set the stage for the search and establishment of policies in aging societies.

At the regional level, in 2007, the Brasilia Declaration was the document that marked the line regarding the protection of the rights of older people, with the firm idea of achieving societies for all ages in Latin America (CEPAL 2011). This would lead to the realization in 2015 of the Inter-American Convention on the Protection of the Human Rights of Older People, a document of paramount importance for the comprehensive well-being of the older population in Latin America. While previous attempts to put the rights of older people on the political agenda were relevant, no document was binding, and they remained only good intentions, unlike the Convention, which stemmed from the conviction "that the adoption of a broad and comprehensive convention will significantly contribute to promoting, protecting, and ensuring the full enjoyment and exercise of the rights of older people, and to fostering active aging in all areas" (OPS 2023, p. 4).

Although the Inter-American Convention became a direct reference for monitoring the Human Rights of older people in the region, Mexico remained on the sidelines as it had not been ratified. Perhaps this was because our country did not have the conditions to comply with what the Convention demanded of the States parties "to adopt, by their constitutional procedures and the provisions of this Convention, legislative or other measures necessary to give effect to such rights and freedoms" (Organización de Estados Americanos [OEA] 2015, p. 3).

It was not until December 2022, published in the Official Gazette of the Federation in January 2023, that the Mexican State would sign the Convention as one of the

States parties, thus committing to comply with the 41 articles it contains, where the general objective is:

> To promote, protect, and ensure the recognition and full enjoyment and exercise, on equal terms, of all human rights and fundamental freedoms of older people, to contribute to their full inclusion, integration, and participation in society. The provisions of this Convention shall not be interpreted as a limitation on broader or additional rights or benefits recognized by international law or the domestic laws of the States Parties in favor of older people. (OEA 2015, p. 3)

Therefore, more than a year after ratifying the Convention, efforts must be made to demonstrate the actions taken in favor of implementing it and to provide this growing age group with the necessary conditions for exercising and monitoring their rights.

As evident from the information presented in the chapter on theoretical positioning, the journey towards recognizing the rights of the older people group through the Convention can be seen in parallel with the MDGs, the SDGs, and the MC, focusing solely on this age group.

6.2 Notes on Demographic Aging

The one we are currently experiencing is known, by the United Nations (UN), as the Decade of Healthy Aging (Organización Mundial de la Salud [OMS] 2021), and it was promoted with the firm conviction that population aging, globally, is undeniable, and therefore efforts should be made to ensure that the aging population reaches old age not only in the best possible health, but also with the best social and economic conditions, as it is expected that by 2050 the number of older people will be double that of today, and by 2100 it will triple. To this situation, we must add the reduction of resources, lack of employment, and the accumulation of diseases, creating a concerning panorama. Thus, it is expected that by 2050, the proportion of people aged 65 and over in the world will be almost the same as that of children under 12 years old; while Latin America and the Caribbean are experiencing a differentiated aging process, it is estimated that it will increase from 9% in 2022 to 19% in 2050 (UN 2022), and it is also noteworthy that the CEPAL has stated that by the year 2040 there will be more people over 65 than children in the subregion (CEPAL 2017), marking a sustained step towards demographic aging.

Thus, Mexico presents as a sociodemographic reality a process of aging, where the increase in the population aged 60 and over has been accentuated in the last 20 years, so that currently it is estimated that 14% of the national population is in this age group, which puts them at a disadvantage in terms of current public policies, since despite it becoming a national agenda item, there are still gaps preventing the full exercise of their rights.

The aging dynamic in Mexico over the last two decades has been as follows:

Projections in our country for 2050 suggest that one in four Mexicans will be over 60 years old, implying that a quarter of the population will be in a stage of life considered unproductive, as the INEGI (2020, p. 2) defines the productive age between 18 and 59 years, a categorization that ends up being discriminatory based on age (Table 6.1).

Now, at the local level, the demographic transition in the State of Mexico has not been much different from that of the country, as it is observed in the second half of the twentieth century, when mortality rates decreased significantly and fertility declined, resulting in a "gradual decrease in the demographic growth rate" (Montoya 2004, p. 7), leading to the demographic aging of the population (Table 6.2).

While the percentage figures for 2020 do not differ much between national and state levels, what should be noted is that in the State of Mexico, there was a difference of 1.6 and 1.8 percentage points from the national data in 2010 and 2000, respectively, which was shortened to more than half in 2020, where it was only 0.7 points. If the aging trend continues, it will align with projections suggesting that by 2040, this population will have doubled, surpassing 3.8 million (Consejo Estatal de Población [COESPO] 2019). Thus, based on the data obtained from 2000 and 2020, it is evident that the population of individuals aged 60 and over has doubled in 20-year periods.

Although this would imply that the population of the State of Mexico will experience full aging by 2040, it is important to question whether society will have the social, economic, and political conditions to achieve comprehensive well-being while respecting the rights of all individuals. Demographic transition is not just a change in the population pyramid; it entails some particularities and consequences, as outlined (Fig. 6.1).

The intention of Fig. 6.1 is to explain that demographic transition is not merely a statistical phenomenon, but rather the current population situation is the result of a historical-social process, determined by its own dynamics, and that in the same sense, it has implications that result, on one hand, in economic consequences (as

Table 6.1 Population aged 60 and over as a percentage of the total population in Mexico

Year	Population in Mexico	Individuals aged 60 years and older	%
2000	97,483,412	6,948,457	7.3
2010	112,336,538	10,109,723	9.1
2020	126,014,024	15,142,976	12

Source Ronzón et al. (2021)

Table 6.2 Population aged 60 and over as a percentage of the total population in State Mexico

Year	Total population in the state of Mexico	Individuals aged 60 years and older in the State of Mexico	%
2000	13,096,686	713,704	5.5
2010	15,175,862	1,137,647	7.5
2020	16,992,418	1,919,454	11.3

Source Own elaboration with information from the Population and Housing Censuses, INEGI (2000, 2010, 2020)

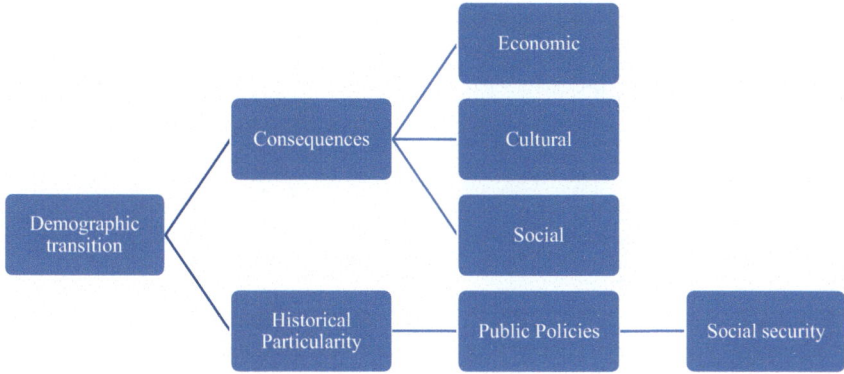

Fig. 6.1 Ramifications of demographic transition. *Source* Own elaboration based on the analysis of demographic transition

the workforce decreases, health costs increase), cultural consequences (one must "learn" to age), social consequences (creating conditions of inclusion for people of all ages), and political consequences (constructing public policies that respond to the needs of an unprecedented reality). Thus, if each of these aspects is not contemplated or addressed, not only are rights violated, but individuals are discriminated against based on age.

6.3 Ageism and Age Discrimination as "New" Forms of Discrimination

Throughout this book, different forms of discrimination have been discussed, some framed by poverty or social conditions, that is, extrinsic characteristics of individuals. However, other forms of discrimination have been presented that relate to intrinsic characteristics, such as age or ethnic background, and when several of these conditions are present in the same person, discriminatory acts can be present daily.

Age has been a condition of discrimination relatively recently because while stereotypes around different stages of life have always existed, they were not a reason or factor for the denial of rights. Even age could play a favorable role for people, in archetypes of youth as valuable or old age as respected. However, social and cultural changes have led to age not being a positive characteristic of the person, but rather, a factor of vulnerability that does not fit in a society tending towards individualism and commercialism.

Thus, the Convention, in its Article 2, defines age discrimination in old age as: "any distinction, exclusion, or restriction based on age that has the purpose or effect of nullifying or restricting the recognition, enjoyment, or exercise, on an equal basis,

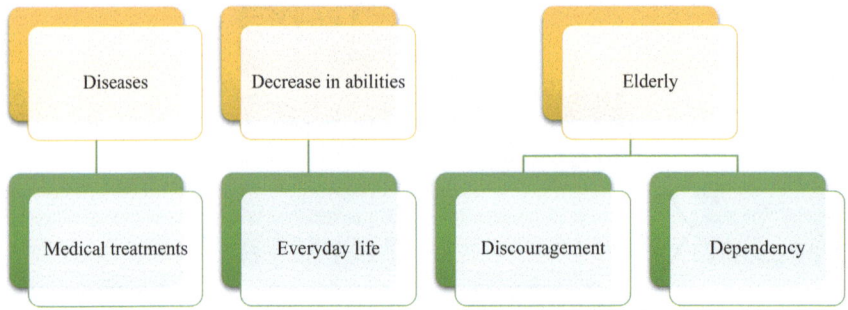

Fig. 6.2 Negative characteristics linked to aging. *Source* Own elaboration

of human rights and fundamental freedoms in the political, economic, social, cultural, or any other sphere of public and private life" (OEA 2015, p. 4). The definition makes it clear that age discrimination in old age can affect all areas of the lives of older people.

Now, the age discrimination that Butler spoke of in 1967 gave way to the concept of ageism, which in Salvarezza (1998) understood as "a prejudiced and discriminatory attitude towards the old" (p. 39). Moreover, it has the particularity that the person, as they age, assumes those same prejudices that they exerted in their youth, about themselves, implying a sort of calamity in growing old (Fig. 6.2).

Figure 6.2 aims to highlight the different characteristics associated with old age as a stage of losses (of health in the first instance) that contribute to the reinforcement of ageism, not only to discriminate against others, the ageing, but even to not want to age, and worse still, to legitimize the discrimination experienced when reaching that stage, and consequently, the violation of rights.

Thus, ageism is the result of the stereotyped social construction of old age, loaded with negative characteristics, mainly derived from physical deterioration.

6.4 From Discriminatory Experiences to the Legitimization of Violated Rights

Discrimination, in general terms, can be experienced daily without even being aware of it, although it may seem hard to believe, since, based on ageism, actions, attitudes, and even omissions are assumed as valid ways of acting towards old age. This is how discriminatory behavior is legitimized in society.

To support the above, the following are excerpts from testimonies of older persons, derived from 16 interviews, in which, even though discrimination was not the main topic, it turned out to be a constant theme around two variables: work or occupation and daily life, in both public and private spaces.

6.4 From Discriminatory Experiences to the Legitimization of Violated Rights

The narratives were collected in different contexts, not only territorial but also social, as some interviews were obtained during activity sessions in different Day Care Centers in Zumpahuacán and Santa Cruz Zinacantepec, rural localities; Tlalnepantla and Naucalpan, which belong to the Valley of Mexico; Toluca and Atlacomulco, important cities for their sociopolitical conditions in the region. Additionally, others were collected in interviews at a permanent stay home of Social Security Institute of the State of Mexico and Municipalities (ISSEMyM) and with members of the "Joyful Friends" group in the Toluca valley, but interviews were also obtained through contact networks. These different realities can be framed within the Montevideo Consensus, given that Priority Measure G, is based on the fact that the territory must be "considered a key element in sustainable development and human coexistence and to reduce territorial inequalities, since these exacerbate economic, social, and environmental inequities, both at the subnational level and among countries" (CEPAL 2015, p. 19), since recognizing the diversity of territories entails recognizing differentiated needs.

Now, regarding work, it is worth noting that almost half of the employed people over 60 years of age in Mexico work on their account (49%), and 70% of the total work in the informal sector (INEGI 2022). Evidence of this is the testimony of Mrs. Toñita, 81-year-old, Mexican, a resident of Tlalnepantla, who responded regarding her occupation:

> Well, I dedicate myself to selling my little trinkets, like these you see here (pointing to a basket with various products she had beside her), "That's what I do. What else can I do at my age? I go out there with the retirees and manage to make enough to eat. (Tlalnepantla 2022)

The testimony of Mrs. Toñita, along with several others, shows that as a growing proportion of the population solidifies their work trajectory in the informal labor market, even from the early stages of their productive life, they will lack an institutional safety net to protect them in the future" (Wong and Aysa 2001).

Regarding work, Evelia, an artisan woman, says:

> Well, I don't back down, because there's always someone trying to pull a fast one. Now, when it came to the stalls for selling our goods (referring to the distribution of spaces for selling her seasonal merchandise), they didn't want to give us a spot. But no, I told them that I have the right because I've been here for many years, and they weren't going to take it away just like that. I need to work. I had eleven children but no one supported me. Sometimes, they give me a few pennies, but not enough to support myself. I keep working because no one gives me anything for free. (83-year-old, Mexican, Toluca 2022)

Mrs. Evelia is evidence of what the ILO maintains regarding the fact that as life stages progress in people's life trajectories, the likelihood of "falling" into informal employment increases in parallel, and in that sense, the possibility is even greater for women (75% compared to 68% of men) (CEPAL and OIT 2018).

Mr. José, a farmer in Zumpahuacán, regarding his current occupation, says:

> Well, right now I don't do much. Sometimes I go to the field, but my children tell me that there is no point. But I do things, I'm still weeding and checking on the corn a bit. I can still do it! Even though my children say I can't (72-year-old, Mexican, Zumpahuacán, 2020).

This testimony is evidence of how economic activity and inactivity have been conceived, as even the ILO recognizes, in these terms, the life cycle in three stages:

(a) Childhood and youth, when individuals undergo education that prepares them for entering the workforce,
(b) Active or working age,
(c) Old age, in which they exit the workforce and live off a pension or assets accumulated in the previous phase (CEPAL and OIT 2018, p. 19).

To these conditions, it should be added that the lower participation of women in the labor market, especially of the aging generations, often was intermittent within the productive stage of their lives. They had to juggle activities as mothers/wives, resulting in a disconnection from formal work systems and a lower accumulation of resources. This leads them to face mere survival in old age through informal work (Zúñiga and García 2008), which likely leads to highly impoverished conditions.

While the ILO itself clarifies that reality is much more complex, these social policies indeed validate collective knowledge, where old age would imply abandoning the activities that older people dedicated most of their lives. In this regard, concerning occupation, life experience can be very different depending on the social environments in which people live, as shown by Mr. Daniel's experience:

> What I want is a job that allows me to contribute to the IMSS [Mexican Institute of Social Security], because I already checked, and they told me that I can qualify for a pension because I worked all my life, and I spent many years at the IMSS (referring to contributions to the institution), but they tell me that I need to start working, and with two years I can retire. That's what I'm looking for, but they just don't give jobs anymore. What I'm seeing is that a friend of mine gives me a job in his business, anything really, the important thing is that I can get into the IMSS. (65-year-old, Mexican, Ecatepec de Morelos 2023)

These testimonies are evidence of what, according to the CEPAL and the ILO, happens in the region, where most older people work for themselves, probably because of age discrimination, in this case, ageism "Ageism hinders older people's access to wage employment and their desire to work independently, leveraging the skills acquired throughout their working lives, to do so under conditions that allow greater flexibility in organizing their work activities and daily lives" (2018, p.18).

Now, regarding the discrimination experienced in their daily lives, it becomes evident both in public spaces (during community interactions) and in private settings (with family and acquaintances). The daily life of older persons involves recreational activities, family gatherings, and caregiving responsibilities, including habits or customs they engage in, either by their own choice or as assigned by social roles.

Conchis, regarding her daily activities, says:

> Well, now I don't do much. Before, I used to go dancing with my husband, but not anymore. Since the pandemic, it's been over for us. It stopped happening, and then when we wanted to go again, we couldn't anymore. They're doing other things there now. They took our spot away… I hope they let us go back because that was part of our life for us. (78-year-old, Mexican, Zinacantepec 2023)

6.4 From Discriminatory Experiences to the Legitimization of Violated Rights

Don Emilio, regarding the activities he performs at the Day Center in Santa Cruz, Zinacantepec, commented:

> I don't know why they have us do these things, what good will this do for me? (showing a raffia flower he was making), but here I am... It's not that I don't like coming here, because I know I have to come. Besides, what else am I going to do at home? Just grow old? But maybe we could do other things. (70-year-old, Mexican 2022)

It is clear that upon reaching old age, differences in life trajectories between men and women become evident, especially among the generations of people who are now ageing, as there was a male tendency to remain more in the labor market, even when they are already retired (Rodríguez and Rossel 2009). However, being a man does not imply not being a victim of discrimination, as Don Emilio perceives that the activities they are assigned to do are not intended for people like him. Today, there is a gender perspective in the integration of groups.

Moreover, retirement is one of the biggest changes experienced by an older man, as it generally leads to a loss of status, reduction in support networks, and decrease or loss of income, implying a redefinition in relationships with everyone around him, as well as in how to face what lies ahead, impacting all levels of a person's life (González 2010). This places him at a disadvantage not previously experienced, having to engage in activities he had not done before, especially because the private space, the home, was not his, his space was the public one, the outside, the workplace, now feeling discriminated against in his own home.

However, these situations can be similar to those of women when the caregiver role persists, even after raising children, assuming responsibilities imposed by their social group. Mrs. Malú, in Naucalpan, speaks about her daily life:

> Now I am enslaved to my granddaughter, something I didn't experience with my children because I always had my things, they went to school and I was fortunate to have someone to help me. But now it's the opposite, I have to lend a hand to my daughter, well, and since she works and her husband too, and they can't afford to pay someone to help with the girl, well, I'm the one who goes up and down with her. I pick her up from school every day and take her home, and if she has classes for the things her mom signs her up for, then I go there with her. But that's how I am, enslaved to my granddaughter! And what's left for me? Well, nothing. (62-year-old, Mexican 2023)

This testimony highlights how aging is perceived as a stage of life in which one must serve others, and their social group, without being recognized as subjects of rights themselves. It's part of a generic and stereotyped view of old age, where the caregiving grandmother must perpetuate that role to be recognized as a member of her family group, thereby compromising her autonomy. Additionally, González and Sánchez (2003) indicate that in old age, family is an important social support network. For example, when grandparents are responsible for caring for grandchildren, they expect to be rewarded with affection, moral support, and financial or in-kind assistance. However, this is not always the case.

In this same vein of autonomy and decision-making in everyday life, Mrs. Ilse says:

> Well, I didn't used to live here; I moved here just because of the pandemic. I used to live in Mexico City; I'd go to Ciudadela, it was nice. But then the pandemic hit, and well, I was alone there, and things got tough. My son, who already lived here with his family, told me it would be better if I moved here because nobody was looking out for me there. So, I didn't have much choice, and here I am... But I miss it there. I used to go downtown and buy things, you know? I'd go to Ciudadela; I love dancing, dancing is so beautiful. Do you see me? I couldn't stop! I even had a dance partner and everything. But, well, it was for my own good, right? But here, there's nothing! We just come here (to the day center) to do crafts and sometimes have conversations. Oh, and yes, there's music too; sometimes they bring in a music and singing teacher, but not always. That's nice. But dancing, no. I'd have to go all the way to Toluca, and that's more complicated than taking the subway. Plus, I don't know anyone there, and it's expensive. Maybe I could go once in a while, but I'm sure my son wouldn't want to. (79-year-old, Mexican, Atlacomulco 2023)

The conditions experienced by Mrs. Ilse illustrate how the lives of older persons must change to adapt to their children's lives. They leave behind their own lives, friendships, and daily routines to integrate into the family life of their descendants. Under these conditions, it can be assumed that when widowed or divorced, the older person becomes more vulnerable, and it is she who possibly changes her place of residence to one of her children's homes. This is evidence of what the Consensus states in its considerations of Chapter C, regarding the importance of addressing old age and aging: "Recognizing that empowering older people is a key dimension for the full respect of their human rights and their full participation in a development model that, to be sustainable, must be inclusive" (CEPAL and NU 2015, p. 10). It is therefore a necessity to empower older persons in their old age.

For single women, especially those who have had informal work trajectories (if any), their situation is particularly complex. This is especially true if we consider the possibility of them having chronic illnesses due to the absence of an institutional safety net that protects them, as well as the reduction of family support networks and exchanges, particularly among the most vulnerable sectors of the population, the impoverished sectors (Villagómez 2010).

> Doña Petra, referring to the support provided among older single women, says:
>
> I have my little friends that I've made here," [referring to the group of older persons at DIFEM- System for the Integral Development of the Family of the State of Mexico], and when one of them doesn't come to the meeting, I go to see her, to check if she's okay. If she's sick, I make my rounds, and I bring her some bread, or a little soup, anything really, because we are alone. (77-year-old, Mexican, Zumpahuacán 2020)

However, it is worth noting that loneliness in rural areas is not exclusive to women. When asked how they manage to obtain resources for food, Mr. Casimiro and Mrs. Leo say:

> Well, just from my daughter-in-law, my son's wife. In the afternoons, she brings me a plate of food almost every day... I accept it when she gives it to me, but if she doesn't bring it, I don't ask... They have their things to do, their children. (Casimiro, 72-year-old, Mexican, Zumpahuacán 2020)
>
> Nope, from whom? From me! If I want to eat, I make something for myself. I always try to have some beans; the ones they give me last, well, until they run out. (Leo, 76-year-old, Mexican, Zumpahuacán 2020)

The breakdown or loss of support networks is alarming, as these are vital for every human being, especially for older persons. Care can be crucial for their well-being, to avoid feeling lonely and falling into depression, which is a recurrent pathology in old age. Additionally, support networks can help improve their self-perception (Kalish 1996). The main support network is the family, as it is the first contact of the human being with society (Bazo 1990).

6.5 Ageism Through Intersectionality

The cases presented lead us to an analysis that must be understood through intersectionality, which allows us to observe the aging sector from the perspective proposed by the MC (CEPAL 2015), considering intrinsic conditions such as social processes, local public policies, urban processes, among others, and extrinsic conditions such as the environment and globalization. Thus, among the agreements of the MC are:

> Promote the development and well-being of people in all territories, without any discrimination, including full access to basic social services, and the equalization of opportunities within cities, between urban and rural areas, among small, medium, and large cities, and between dispersed populations and those living in small rural settlements (p. 20).

This is a task that is not evident in the testimonies; the actions of the state fail to have this perspective, and older people have different problems because the experiences of aging are different in the State of Mexico. Intersectionality can provide a series of tools to identify the violation of the rights of older people even though they are not aware, as when activities in Day Care Centers do not meet their needs, preferences, or gender perspective, their rights are violated, denying them decision-making power and inclusion in applicable public policies for this sector.

The reality experienced by older persons in the State of Mexico demonstrates a failure to comply with the agreements of the Montevideo Consensus, as outlined in number 18, which states that states will develop policies at all levels to guarantee:

> The quality of life, the development of potentiality, and the full participation of older people, addressing the need for stimulation (intellectual, emotional, and physical) and considering the different situations of men and women, with special emphasis on the groups most susceptible to discrimination (older people with disabilities, those lacking economic resources and/or pension coverage, and older people living alone and/or without support networks). (CEPAL 2015, p. 21)

Therefore, discrimination against older persons is reflected in poorly designed social programs, and unfocused activities imposed from above, placing them in a situation of intersectional discrimination (Fig. 6.3).

The purpose of Fig. 6.3 is to highlight how both a man and a woman, in old age, face conditions that put their rights at risk. In this regard, the United Nations has pointed out that thousands of older persons in Latin America experience situations of discrimination and abandonment "which worsens when it comes to women,

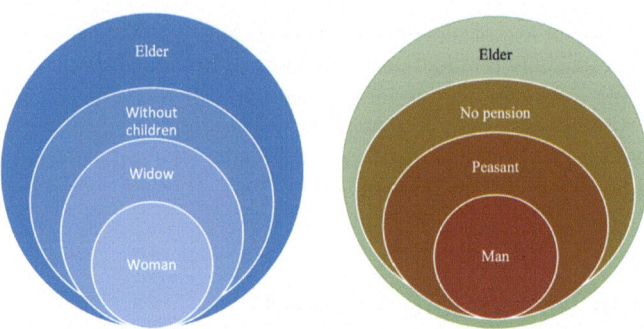

Fig. 6.3 Examples of intersectional discrimination for a man and a woman. *Source* Own elaboration based on testimonies from older persons

Afro-descendants, indigenous people, refugees, displaced persons, stateless persons, LGBTIQA+ individuals, and people with disabilities, as they are exposed to multiple forms of discrimination" (OPS 2023, p. 6).

6.6 Conclusion

To close this chapter, several points should be highlighted. The current sociodemographic conditions in the State of Mexico have never been seen before, leading to accelerated aging, which complicates not only the protection of rights but also interpersonal relationships within communities.

The situation becomes even more severe for older persons who encounter an undeniable lack of job opportunities, family rejection/abandonment, denial of health services, or indifference and mistreatment in public spaces, evidencing a growing discrimination and violation of their rights daily. This discrimination has been legitimized by the older persons themselves, perhaps unknowingly, allowing the violation of their rights through discriminatory acts such as: being unable to make decisions about their own lives, prioritizing the well-being of others (children, grandchildren) to avoid "bothering" them, with the conviction that others come before themselves.

Thus, the violation of the rights of older persons is perpetuated through discrimination, ageism, and elder abuse, which have been legitimized by those experiencing old age because they replicated them in earlier stages of life, becoming a cyclical constant across different generations.

Moreover, as observed throughout the chapter, the perpetuation of stereotypes about old age and aging turns back on the same aging society, placing the population in constant vulnerability, partly due to a lack of knowledge of their rights and partly due to the state's failure to monitor those rights.

It was also evident that in Mexico, various laws establish the guarantees that older adults must have to enjoy an adequate standard of living. Additionally, in April 2024, the State of Mexico reformed the Law for Older Adults (Poder Ejecutivo del Estado de México 2024) with a geriatric-gerontological vision, recognizing the particular conditions of this age group that must be addressed by the state and in line with various international treaties that address the rights of older adults. The task now is to demand those rights.

In summary, discrimination against older adults, not only in Mexico but also in various parts of the world, remains a structural and complex problem (Rodríguez 2023) that requires a comprehensive approach to address its causes and consequences. It is essential to promote sociocultural and political changes based on the inclusion and recognition of all age groups to ensure a fairer and more equitable society for all individuals.

References

Bazo, M.T. (1990). *La sociedad anciana*. Siglo XXI
Comisión Económica para América Latina y el Caribe. (2011). *Declaración de Brasilia*. UNFPA. https://repositorio.cepal.org/entities/publication/50e450df-000c-40a3-9a03-642c5d7531b7
Comisión Económica para América Latina y el Caribe. (2015). *Guía Operacional para la Implementación y el Seguimiento del Consenso de Montevideo Sobre Población y Desarrollo*. CEPAL. https://www.cepal.org/es/publicaciones/38935-guia-operacional-la-implementacion-seguimiento-consenso-montevideo-poblacion
Comisión Económica para América Latina y el Caribe. (2017). *Cuarta Conferencia Regional Intergubernamental sobre Envejecimiento y Derechos de las Personas Mayores*. CEPAL. https://www.cepal.org/es/notas/cuarta-conferencia-regional-intergubernamental-envejecimiento-derechos-personas-mayores
Comisión Económica para América Latina y el Caribe, Naciones Unidas. (2015). *Guía Operacional para la Implementación y el Seguimiento del Consenso de Montevideo sobre Población y Desarrollo*. CEPAL, NU. https://repositorio.cepal.org/entities/publication/b7666bc8-b86d-4808-ba0b-23ab16e97c0c
Comisión Económica para América Latina y el Caribe, Organización Internacional del Trabajo. (2018). La inserción laboral de las personas mayores: necesidades y opciones. *Coyuntura Laboral en América Latina y el Caribe*, (18). https://repositorio.cepal.org/server/api/core/bitstreams/f4a18703-baa6-4f08-a86e-7c665ff42b13/content
Consejo Estatal de Población. (2019). *Envejecimiento Demográfico*. Gobierno del Estado de México. https://coespo.edomex.gob.mx/sites/coespo.edomex.gob.mx/files/files/2019/ENVEJECIMIENTO%20demografico.pdf
González, C., y Sánchez, J.J. (2003). Efectos de un programa cognitivo-conductual para mejorar la calidad de vida en adultos mayores. *Revista Mexicana de Psicología*, 20 (1), pp. 43-58. https://psycnet.apa.org/record/2004-10383-004
González, R. (2010). Calidad de vida en el adulto mayor. En *Envejecimiento Humano. Una visión transdisciplinaria* (pp. 365–377). Instituto de Geriatría. https://www.gob.mx/inger/documentos/envejecimiento-humano-una-vision-transdisciplinaria
Instituto Nacional de Estadística y Geografía. (2010). *Censo de Población y Vivienda 2010*. INEGI. https://www.inegi.org.mx/programas/ccpv/2010/

Instituto Nacional de Estadística y Geografía. (2020). *Estadísticas a propósito del Día Mundial de la Población (11 De Julio).* INEGI. https://www.inegi.org.mx/contenidos/saladeprensa/aproposito/2020/Poblacion2020_Nal.pdf

Instituto Nacional de Estadística y Geografía. (2022). *Encuesta Nacional de Ocupación y Empleo, Nueva Edición. Primer trimestre 2022.* INEGI. https://www.inegi.org.mx/programas/enoe/15ymas/

Kalish, R. (1996). *La vejez: perspectiva sobre el desarrollo humano.* Pirámide.

Montoya, J. (2004). Los retos demográficos en el Estado de México. *Papeles de Población,* nueva época, 10 (40), 25–29. https://rppoblacion.uaemex.mx/article/view/8754/7461

Naciones Unidas. (1991). Principios de las Naciones Unidas en favor de las personas de edad. *Resolución 46/91 de la Asamblea General de las Naciones Unidas del 16 de diciembre.* NU. https://www.un.org/development/desa/ageing/resources/international-year-of-older-persons-1999/principles/los-principios-de-las-naciones-unidas-en-favor-de-las-personas-de-edad.html

Naciones Unidas. (2010). Principios de las Naciones Unidas en favor de las personas de edad. https://www.un.org/development/desa/ageing/resources/international-year-of-older-persons-1999/principles/los-principios-de-las-naciones-unidas-en-favor-de-las-personas-de-edad.html

Organización de los Estados Americanos. (2015). *Convención Interamericana sobre la Protección de los Derechos Humanos de las Personas Mayores. Washington.* OAS. https://www.oas.org/es/sla/ddi/docs/tratados_multilaterales_interamericanos_a-70_derechos_humanos_personas_mayores.pdf

Organización Mundial de la Salud. (2021). *Década del envejecimiento saludable: informe de referencia. Resumen.* OMS. https://iris.who.int/handle/10665/350938

Organización Panamericana de la Salud. (2023). *La Convención Interamericana sobre la Protección de los Derechos Humanos de las Personas Mayores como herramienta para promover la Década del Envejecimiento Saludable.* OPS. https://iris.paho.org/handle/10665.2/57353

Poder Ejecutivo del Estado de México. (2024). *Decreto por el que se cambia la denominación de la Ley del Adulto Mayor del Estado de México, por la Ley de las Personas Adultas Mayores del Estado de México, y por el que se reforma y adicionan diversas disposiciones de la misma.* https://legislacion.edomex.gob.mx/sites/legislacion.edomex.gob.mx/files/files/pdf/gct/2024/julio/jul171/jul171a.pdf

Rodríguez, F. y Rossel C. (Coords.) (2009). *Panorama de la vejez en Uruguay.* UCU-UNFPA.

Rodríguez, J. (2023). *Una teoría de la discriminación.* Universidad Autónoma Metropolitana-Iztapalapa.

Ronzón, Z., Méndez, A. y Jardón, A. (2021). El derecho a los cuidados de las personas mayores, una necesidad del sistema de Salud en México. *La Situación Demográfica de México 2021.* CONAPO. https://www.gob.mx/conapo/documentos/la-situacion-demografica-de-mexico-2021

Salvarezza, L. (Comp.) (1998). *La Vejez. Una mirada gerontológica actual.* Editorial Paidós.

United Nations. (2022). *World population prospects summary of results* [Informe UN DESA/POP/2022/TR/ NO. 3]. Departamento de Asuntos Sociales y Económicos. https://www.un.org/development/desa/pd/content/World-Population-Prospects-2022

Villagómez, P. (2010). El envejecimiento demográfico en México: niveles, tendencias y reflexiones en torno a la población de adultos mayores. En *Envejecimiento Humano. Una visión transdisciplinaria* (pp. 305–314). Instituto de Geriatría.

Wong R. y Aysa, L. (2001). Envejecimiento y salud en México: un enfoque integrado. *Estudios Demográficos y Urbanos,* 16(3), 519–544. https://doi.org/10.24201/edu.v16i3.1107

Zúñiga, E. y García, J. (2008). El envejecimiento demográfico en México. Principales tendencias y características. *La situación demográfica de México 2008.* https://www.gob.mx/conapo/documentos/situacion-demografica-de-mexico-2008

Chapter 7
Conclusions and Recommendations

The book has argued that discrimination is an act based on power relations, where certain groups wield discriminatory actions to maintain and reproduce conditions that exert control over those whose rights are violated.

This perspective has underpinned the analysis of the different realities presented throughout this work, with the ultimate goal of contributing to the construction of more equitable, peaceful, tolerant societies with equal rights, thus achieving social, sustainable, and inclusive development.

What is needed, then, is a perspective of equal rights and equity of conditions to ensure that all individuals, with special attention to socially vulnerable groups, enjoy, not only in the State of Mexico but throughout the country, a society free of discrimination. This will be possible through slow but steady efforts originating from comprehensive public policies and strengthened within schools among the new generations, as well as being instilled within the dynamics of households) (Fig. 7.1).

The previous scheme aims to make evident how the book envisions the possibility and the path to build societies that live without discrimination, where the cornerstone is the surveillance by the State and civil society of the Human Rights of all individuals, integrating the different international efforts that this book has taken to create conditions that build communities where their culture does not reproduce discriminatory acts and achieve, therefore, the sustainable development of peaceful and inclusive societies.

7.1 The Proposals

In this book, various expressions of discrimination have been highlighted in: (a) international migrant persons, (b) indigenous women, (c) youth, and (d) older persons. Based on this, the need to generate recommendations that contribute to the construction of peaceful and inclusive societies in Mexico is recognized, which can

Fig. 7.1 Elimination of violence from a rights perspective. *Source* Own elaboration

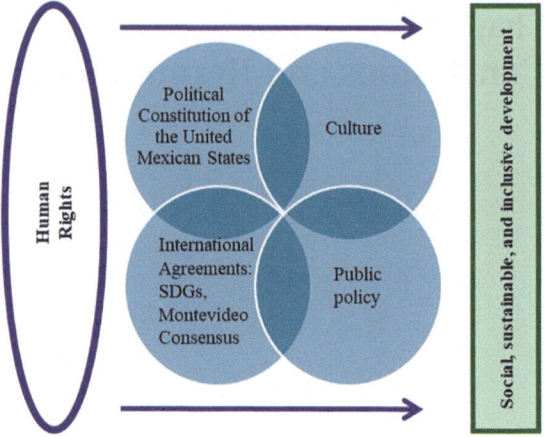

and should go hand in hand with the generation and implementation of actions aimed at monitoring the fulfillment of the rights of all people, as well as achieving the SDGs for the 2030 agenda and the Montevideo Consensus.

Eradicating discrimination thus becomes a complex issue, which requires being approached with a procedural, long-term perspective involving different actors. Hence, proposals are presented for each population group to consider their characteristics and particular contexts, but also to emphasize the commitment required from various entities to eradicate discrimination in these and other areas of society.

7.1.1 International Migrant Persons

For cases involving individuals in contexts of human mobility, it is recommended, firstly, to support and promote diagnostic studies and specific experiences that allow the complexity of the migration context in the State of Mexico to be understood, as well as to make these populations and their specific needs visible. These studies, as previously stated, constitute inputs for the public agenda, providing tools to rethink migration policy from a comprehensive perspective, particularly given its shortcomings. This recommendation needs to be implemented through strategies supported by human rights education and cultural diversity, as redirecting anti-immigrant stances, hate speech, and disdainful attitudes towards this population depends on it.

Specifically, the following recommendations are listed:

For the society of the State of Mexico:

- Promote, through formal and informal educational processes, awareness of human rights education, aiming to advance the sensitization of society and the adoption of attitudes of acceptance, inclusion, and non-discrimination towards this social group.

For public officials and the government system:

- Advocate for the enactment of regulations on interculturality and inclusion, as well as recognizing these principles in existing legal instruments.
- Promote the allocation of financial, human, and material resources for managing migrations and mobilities, adhering to the principle of non-discrimination.
- Build alliances with other stakeholders, particularly with international organizations such as UNHCR, IOM, and ILO.
- Foster links with the government of Mexico City, to build synergies to address these issues in a coordinated manner for the benefit of the migrant and refugee population residing in the municipalities of the metropolitan area.
- Promote training and education to recognize the complexity of the migration context in the region, the involved populations, and their rights.
- Drive processes of awareness-raising and the development of intercultural competencies among public servants, to prevent discriminatory treatment towards the population accessing institutional spaces.

For civil society organizations:

- Promote the creation and establishment of new spaces to address the immediate needs of migrant populations in transit and in other mobility situations.
- Coordinate and establish links with the public administration and other organizations to expand the current support network for migrants in this region.

In summary, these recommendations suggest that addressing discrimination against international migrant populations and in other contexts of human mobility requires cooperation and coordinated, specific, and punctual work among the various actors involved in accompanying and protecting this social group.

7.1.2 Indigenous Women

Regarding actions aimed at eradicating the multiple discriminations faced by indigenous women who work as domestic workers, as analyzed in this book, it is possible to consider at least two dimensions: discrimination based on ethnic origin and in terms of labor.

Regarding the first dimension:

- It is essential for federal and state public institutions to strengthen efforts aimed at enforcing anti-discriminatory regulations towards indigenous people in various

spheres through specific programs, including a gender perspective to recognize the specific discriminations faced by indigenous women.
- These efforts should involve the participation of diverse indigenous groups (including women) in the design and management of public actions concerning them, thus avoiding paternalistic approaches.
- Due to the relational nature of discrimination, it is also necessary to promote a culture of respect and appreciation for cultural diversity expressed in indigenous peoples, as well as the eradication of racist prejudices towards indigenous people in general and indigenous women in particular. This also entails fostering a culture of reporting and addressing complaints.

Regarding the labor conditions of women who work as domestic workers:

- Institutional actions must be strengthened to enforce the International Labour Organization's Convention 189 and promote decent work and dignified treatment in this sector, both in Mexico City and throughout the rest of Mexico, as it is a widespread issue.
- Mechanisms should be strengthened to ensure full access to social security, the use of written contracts, and the establishment of fair wages for this labor sector.
- In cultural terms, efforts should be made to valorize domestic work, raise awareness among employers, eradicate classist and racist prejudices, and implement effective actions to formalize the employment relationship and prevent discriminatory practices so that the workplace (the employer's home) is safe and free from violence.

All the aforementioned actions must include the participation of domestic workers in identifying discriminatory practices they face and their labor-related needs through dialogue with activists, associations, and domestic workers in various contexts, ensuring respect for the language, customs, and cultural particularities of indigenous women.

7.1.3 Young Population

For young people who study and work, the need to create appropriate policies and allocate necessary resources to transform an activity aimed at meeting needs into a positive strategy involving greater knowledge acquisition, linking theory with practice, and generating more employment opportunities is highlighted. To achieve this, the following recommendations are made.

For the international agenda regarding labor matters:

- Develop guidelines that specify exceptions that prohibit the right to equal opportunities and treatment, to avoid covering up discriminatory acts.
- Develop specific indicators for various groups of young people, to distinguish heterogeneities regarding their sociodemographic characteristics, thus providing sensitive, differentiated, and specific attention.

7.1 The Proposals

For the Mexican government:

- Ensure the free exercise of labor rights.
- Implement measures to prevent discrimination from an intersectional and human rights perspective.
- Establish channels of information about labor rights, both for young people and for public officials, employers, and society in general.
- Regulate the hiring of young people, especially when conditions for full-time work are not met.
- Manage intersectoral programs that support labor rights, such as in the fields of education, health, care, economy, and transportation, among others.

For employers and society in general:

- Report acts that threaten or violate labor rights.
- Promote jobs that ensure the full inclusion of young people in the labor market.
- Encourage employability strategies that include more training focused on job positions.
- Recognize the diversity of jobs that young people engage in and ensure that their labor rights are respected.

In summary, the aim is to promote a culture of respect for the free exercise of rights of young individuals. Labor regulations should align with the protection of human rights, not at the expense of it. Lack of experience should not be punished with low wages or the elimination of statutory benefits. Simultaneously engaging in various significant activities in the lives of young people should not be an excuse to undermine their work.

7.1.4 Older People

To address the needs of the population aged 60 and older, several proposals are made not only for monitoring and restoring their rights but also for the development of current and appropriate public policies regarding population aging in Mexico. These proposals aim to respond to the sociocultural diversity of the nation:

For the fulfillment of the international agenda:

- Comply with the Inter-American Convention on Protecting the Human Rights of Older Persons and use it as a guide for monitoring and enforcing the rights of the older population.
- Adhere to Article 25 of the Universal Declaration of Human Rights, which seeks to ensure that all individuals have an adequate standard of living, informing the development of public policies.
- Integrate a gerontogeriatric perspective into the implementation of actions to achieve the SDGs, to establish lines of action for an aging society.

To combat the impoverishment of the population aged 60 and older by the Mexican State:

- Restructuring of the Law on the Rights of Older Persons, where the focus is not only on the vulnerability of this social group but also on monitoring the implementation of their social, economic, political, and health rights.
- Changing existing welfare-based public policies to policies aimed at solving the problems faced by the ageing population today.
- Creation of social programs in the labor area to promote the integration of the aging population into the formal labor market and provide them with social security benefits.
- Establishment of a pension system that allows for the incorporation of retirement funds for individuals who do not work in the formal sector, with support from the State.
- Considering ageing care as a right, where responsibility is not solely on the family but is equitably distributed among the State, family, and society.
- Health care should be comprehensive, addressing physical, mental, and social aspects. There is an urgent need to review established policies for IMSS well-being, including the provision of care for diseases and conditions prevalent in old age with a gerontogeriatric perspective.

For society in general:

- Deconstruct and reconstruct the culture of old age and aging, shifting towards a social understanding of old age as a stage of life that can be active, productive, and functional. This reconstruction should be guided by public policies that involve the entire population, making it comprehensive and cross-cutting.

The previous recommendations aim to build a society for all ages, free from discrimination, with the monitoring of rights, and preparing society.

7.2 The Contributions

According to the research working group, this book offers five main contributions to the study and analysis of discrimination. Firstly, it discusses the concept from its simplest construction to a more elaborate and complex understanding, emphasizing the need to incorporate a perspective from human rights and intersectionality. This broader perspective helps to visualize the various expressions of discrimination in society and generate focused actions to efficiently eradicate it.

The second contribution focuses on the process of including discrimination in Mexico's public agenda. Before 2000, the existence of this social issue in the country was denied. However, the formation of the Citizen Commission for Studies against Discrimination was crucial for subsequent legislative processes and anti-discrimination policy formation. Currently, Article 1 of the Mexican Constitution is a top-level legal instrument that recognizes human rights and grounds for discrimination that undermine the dignity and rights of individuals. Additionally, the Federal Law and the role of CONAPRED in federal anti-discrimination policy are highlighted. Over two decades, substantial progress has been made in this field, but it is recognized that more needs to be done to eradicate discrimination in the country.

The third contribution is the assertion of the prevalence of discrimination in Mexico for the four analyzed population groups. The National Survey of Discrimination (ENADIS) provided a clearer and more precise context on the perception of rights, issues, prejudices, stereotypes, denial of rights, situations, motives, and areas of discrimination. This reveals the heterogeneity of discriminatory expressions experienced in various contexts and due to overlapping reasons, making intersectionality explicit.

The fourth contribution stems from the analytical depth of the four population groups with meeting points in the State of Mexico: international migrants, indigenous women, young people, and the ageing. Through direct dialogues, multiple experiences of those who have suffered discriminatory acts are recovered and analyzed. This allows for reflection on the prevalence and persistence of this social issue, which is often rooted, normalized, and detrimental to the lives of individuals and broader environments.

The fifth contribution of this book lies in the analytical complexity of discrimination. The various researches that underpin this book were conducted from the field of Social Sciences and Humanities, allowing for a more comprehensive view provided by inter-, intra-, and multidisciplinarity. Taking a stance from a single discipline would imply an analytical bias, making dialogue between various scientific areas of knowledge indispensable for this product.

On the other hand, while recognizing the contributions of this work, some limitations are also detected. Firstly, due to the focus on Social Sciences and Humanities, other areas of knowledge that could enrich the analysis were not deeply explored. For example, when discussing perceptions of discrimination, psychological perspectives could have been included to examine psychosocial impacts, and closer collaboration with health sectors could have been established to identify the implications of discrimination on individuals. Another limitation is related to the selection of individuals who contributed their experiences. Individuals tasked with guaranteeing, safeguarding, and upholding the rights of the population were not considered, to give space to testimonies of individuals who suffer discriminatory acts and, consequently, the violation of their fundamental rights.

However, these limitations provide opportunities for further research and potential lines of inquiry that delve deeper into and emphasize the need to address discrimination in all its forms, expressions, and contexts. This importance lies in the search for alternatives to promote an anti-discriminatory culture, where the rights of all individuals are upheld, aiming to build peaceful and inclusive societies.

In conclusion, the first and most pressing recommendation is to continue generating information on the different forms of discrimination experienced daily by multiple population groups in the national territory, to directly and urgently impact the sensitization and training of public servants, employers, and the general population regarding inclusion and non-discrimination towards groups whose rights are habitually violated. This is essential to achieve the benefits of equality, inclusion, and the full enjoyment of rights for all individuals.

On the Institution

The Autonomous University of the State of Mexico (UAEMéx)

Mission

The Autonomous University of the State of Mexico is a public institution with full autonomy, which provides high school and higher education with high quality standards, equity and inclusion in different regions of the state. It fulfills its commitment to educate ethical, integral, critical, supportive and reflective people with an emphasis on peace building, possessing a solid preparation and universal awareness and committed to gender equality, the promotion and defense of human rights, justice, democracy, values, sustainable development and the common good. It contributes to interpret the world around us, carries out activities of creation, dissemination and extension of humanistic, scientific and technological knowledge and of the different cultural, artistic and physical culture manifestations to contribute to the development and continuous improvement of the level of welfare of society. It promotes transparency and timely accountability under the auspices of an efficient, responsible and responsive management, at the service of society.

University Administration 2021–2025

Website: www.uaemex.mx/

On the Authors

Ana Elizabeth Jardón Hernández holds a Ph.D. in Social Sciences from El Colegio de Michoacán, a Master's in Regional Development from El Colegio de la Frontera Norte, and a Bachelor's in Territorial Planning from the Autonomous University of the State of Mexico (UAEMéx). She is currently a Professor-Researcher at the Center for Research in Social Sciences and Humanities at UAEMéx. Dr. Jardón Hernández is a Level I member of the National System of Researchers, recognized with the Desirable Profile by the Secretariat of Public Education and is part of the UNESCO Chair "Vulnerability and Social Inclusion." Her research focuses on migratory dynamics and human mobility between Mexico and the United States, and she has numerous national and international publications in this field.

Address: Ana Elizabeth Jardón Hernández, Center for Research in Social Sciences and Humanities, Autonomous University of the State of Mexico (UAEMex), Toluca, State of Mexico, Mexico,

Email: aejardonh@uaemex.mx

María Verónica Murguía Salas is a Full-Time Professor at the Center for Research in Social Sciences and Humanities at the Autonomous University of the State of Mexico (UAEMéx). She is a Level I member of the National System of Researchers and holds PRODEP recognition. Dr. Murguía Salas earned a Ph.D. in Social Anthropology from the Universidad Iberoamericana, a Master's in Population and Development from FLACSO-Mexico, and a Bachelor's in Social Anthropology from UAEMéx. With 13 years of experience teaching at the undergraduate and graduate levels, she has taught courses on qualitative and quantitative methodologies, epistemology, demography, and social theory, among others. Her national and international publications focus on her research lines: Anthropology of Labor, Labor Sociodemography, and Youth Studies.

Address: María Verónica Murguía Salas, Center for Research in Social Sciences and Humanities, Autonomous University of the State of Mexico (UAEMex), Toluca, State of Mexico, Mexico

Email: mvmurguias@uaemex.mx

Itzel Hernández Lara is a sociologist from the National Autonomous University of Mexico (UNAM) and holds a Master's in Regional Studies from the Instituto Mora. She completed a Ph.D. in Social Science with a specialization in Sociology at El Colegio de México. Dr. Hernández Lara is currently a professor-researcher at

UAEMéx, affiliated with the Faculty of Political and Social Sciences. Her research lines include labor mobility from rural communities, migration, family life, gender relations, and the sociology of emotions. She is a member of the National System of Researchers (Mexico).

Address: Itzel Hernández Lara, Faculty of Political and Social Science, Autonomous University of the State of Mexico (UAEMex), Toluca, State of Mexico, Mexico

Email: ihernandezl@uaemex.mx

Zoraida Ronzón Hernández holds a Ph.D. in Social Anthropology from CIESAS-Mexico City and has been a Professor-Researcher at the Center for Research in Social Sciences and Humanities at UAEMéx since 2010. She has received SEP-PRODEP recognition and is a Level I member of the National System of Researchers. Dr. Ronzón Hernández has contributed to national policy discussions, such as working groups with the National Population Council on the creation of the National Population Policy and the revision of the Montevideo Consensus. She has collaborated with UNFPA Mexico and the governments of Mexico City, the State of Mexico, Guanajuato, and Veracruz on proposals and policies addressing older adults. She participated in updating the Rights of Older Adults Law in the State of Mexico in 2024. Her research lines focus on age groups, particularly aging and old age, as well as human mobility from a gender perspective.

Address: Zoraida Ronzón Hernández, Center for Research in Social Sciences and Humanities, Autonomous University of the State of Mexico (UAEMex)

Toluca, State of Mexico. Mexico

Email: zronzonh@uaemex.mx

Bibliography

Agencia de la ONU para los Refugiados. (n.d.). *Buena práctica 13: no discriminación por situación migratoria*. https://acnur.org/fileadmin/Documentos/Proteccion/Buenas_Practicas/9219.pdf

Agencia de la ONU para los Refugiados. (2023). *Caminando hacia la integración. Principales resultados*. ACNUR México. https://www.acnur.org/mx/media/caminando-hacia-la-integracion-2022-principales-resultados-de-acnur-mexico

Alarcón, R., Cruz, R., Díaz, A., González, G., Izquierdo, A., Yrizar, G. y Zenteno, R. (2009). La crisis financiera en Estados Unidos y su impacto en la migración mexicana. *Migraciones Internacionales*, 5(16), 193–210. https://migracionesinternacionales.colef.mx/index.php/migracionesinternacionales/article/view/1108

Arizpe, L. (1978). *Migración, etnicismo y cambio económico (un estudio sobre migrantes campesinos a la Ciudad de México)*. El Colegio de México.

Arredondo, F. (2010). *Personas físicas nacionales y extranjeras. Régimen jurídico* (2ª. ed.). Colección Colegio de Notarios del Distrito Federal.

Azuela, M. (n.d.). *Manual de buenas prácticas para empleadoras y empleadores justos*. Hogar Justo Hogar. https://caceh.org.mx/wp-content/uploads/2023/12/Manual-de-buenas-practicas-para-empeladores-y-empleadoras-justos.pdf

Banco Mundial. (2018, 19 de septiembre). *Según el Banco Mundial, la pobreza extrema a nivel mundial continúa disminuyendo, aunque a un ritmo más lento* [Comunicado de Prensa N.º 2019/030/DEC-GPV]. https://www.bancomundial.org/es/news/press-release/2018/09/19/decline-of-global-extreme-poverty-continues-but-has-slowed-world-bank

Bautista, M. (2012). Hacer algo distinto para erradicar la discriminación y explotación hacia las trabajadoras del hogar. En Consejo Nacional para Prevenir la Discriminación (Comp.), *Dos mundos bajo el mismo techo. Trabajo del hogar y no discriminación* (pp. 29–38). CONAPRED.

Bautista, M. (2022). *Trabajo del hogar: invisible pero necesario*. CACEH. https://caceh.org.mx/trabajo-del-hogar-invisible-pero-necesario/

Bazo, M.T. (1990). *La sociedad anciana*. Siglo XXI

BBVA México y CONAPO. (2023). *Anuario de Migración y Remesas*. https://www.bbvaresearch.com/wp-content/uploads/2024/03/Anuario_Migracion_y_Remesas_2023.pdf

BBVA Research. (2020). *Mapa de casas del migrante, albergues y comedores en las principales rutas de migración por México*. https://www.bbvaresearch.com/wp-content/uploads/2020/02/Mapa_2020_Albergues_Migrantes_Portable.pdf

Benhumea, L. (2019). El pacto por México: una reflexión sobre el sistema precario de salud Mexicano. *SAPIENTIAE: Revista de Ciencias Sociais, Humanas e Engenharias, 5*(1), 5–30.
Bensusán, G. (2010). Las reformas laborales y el corporativismo mexicano: alternativas en Europa y América Latina. En I. Bizberg (Ed.), *México en el espejo latinoamericano ¿Democracia o crisis?* (pp. 297–358). El Colegio de México, Fundación Konrad Adenauer.
Bucio, R. (2012). Presentación. El desafío de profundizar en el conocimiento de la discriminación para atacar sus raíces. En R. de la Madrid (Coord.), *Reporte sobre la discriminación en México 2012. Introducción general* (pp. 9–11). CIDE, Consejo Nacional para Prevenir la Discriminación.
Bustamante, J. (2002). Immigrants' Vulnerability as Subjects of Human Rights. *International Migration Review, 36*(2), 333–354. http://www.jstor.org/stable/4149456
Bustamante, L., Ayllón, S. y Escanés, G. (2018). Abordando la trayectoria universitaria desde el pensamiento complejo. *Praxis Educativa, 22*(3). https://cerac.unlpam.edu.ar/index.php/praxis/article/view/2648/3217
Cámara de Diputados. (2003). *Ley Federal para Prevenir y Eliminar la Discriminación*. DOF 19-01-2023. https://www.diputados.gob.mx/LeyesBiblio/pdf/LFPED.pdf
Cámara de Diputados del H. Congreso de la Unión (2024). Ley de Migración. Nueva Ley publicada en el Diario Oficial de la Federación el 25 de mayo de 2011. https://www.diputados.gob.mx/LeyesBiblio/pdf/LMigra.pdf
Centro de Investigación en Política Pública. (2022). *El desempeño del mercado laboral mexicano: potencia sin aprovechar*. IMCO. https://imco.org.mx/el-desempeno-del-mercado-laboral-mexicano-potencial-sin-aprovechar/#:~:text=En%20total%2C%20el%20potencial%20no,hombres%20tambi%C3%A9n%20tienen%20empleo%20insuficiente
Centro Latinoamericano y Caribeño de Demografía. (2010). *Impactos de la crisis económica en la migración y el desarrollo. Respuestas de política y programas en Iberoamérica*. CELADE. https://kmhub.iom.int/sites/default/files/impactos_de_la_crisis_economica_en_la_migracion_y_el_desarrollo_0.pdf
Centro Nacional para la Capacitación Profesional y Liderazgo de las Empleadas del Hogar A.C. (n.d.). *Identidad cultural y no discriminación de las empleadas del hogar*. CACEH. https://caceh.org.mx/wp-content/uploads/2018/12/Identidad-cultural.pdf
Centro para la Justicia y el Derecho Internacional. (2019, 27 de septiembre). *México: denuncias ante CIDH violación sistemática de derechos humanos contra migrantes* [Comunicado de prensa]. https://cejil.org/comunicado-de-prensa/mexico-denuncian-ante-cidh-violacion-sistematica-de-derechos-humanos-contra-migrantes/
Chacaltana, J., Dema, G. y Ruiz, C. (2018). El futuro del trabajo que queremos. La voz de los jóvenes y diferentes miradas desde América Latina y el Caribe. *Perfiles educativos*, XL(159), 194–210. https://perfileseducativos.unam.mx/iisue_pe/index.php/perfiles/article/view/58775
Comisión Económica para América Latina y el Caribe. (2011). *Declaración de Brasilia*. UNFPA. https://repositorio.cepal.org/entities/publication/50e450df-000c-40a3-9a03-642c5d7531b7
Comisión Económica para América Latina y el Caribe. (2013). *Consenso de Montevideo sobre población y desarrollo*. Comisión Económica para América Latina y el Caribe. https://www.cepal.org/es/publicaciones/21835-consenso-montevideo-poblacion-desarrollo
Comisión Económica para América Latina y el Caribe. (2015). *Guía Operacional para la Implementación y el Seguimiento del Consenso de Montevideo Sobre Población y Desarrollo*. CEPAL. https://www.cepal.org/es/publicaciones/38935-guia-operacional-la-implementacion-seguimiento-consenso-montevideo-poblacion
Comisión Económica para América Latina y el Caribe. (2017). *Cuarta Conferencia Regional Intergubernamental sobre Envejecimiento y Derechos de las Personas Mayores*. CEPAL. https://www.cepal.org/es/notas/cuarta-conferencia-regional-intergubernamental-envejecimiento-derechos-personas-mayores
Comisión Económica para América Latina y el Caribe, Naciones Unidas. (2015). *Guía Operacional para la Implementación y el Seguimiento del Consenso de Montevideo sobre Población*

y Desarrollo. CEPAL, NU. https://repositorio.cepal.org/entities/publication/b7666bc8-b86d-4808-ba0b-23ab16e97c0c

Comisión Económica para América Latina y el Caribe, Organización Internacional del Trabajo. (2018). La inserción laboral de las personas mayores: necesidades y opciones. *Coyuntura Laboral en América Latina y el Caribe*, (18). https://repositorio.cepal.org/server/api/core/bitstreams/f4a18703-baa6-4f08-a86e-7c665ff42b13/content

Comisión Mexicana de Ayuda a Refugiados. (2023). *La COMAR en números*. COMAR. https://www.gob.mx/comar/articulos/la-comar-en-numeros-327441?idiom=es

Comisión Nacional de los Derechos Humanos. (2012). *La discriminación y el derecho a la no discriminación*. CNDH. https://www.cndh.org.mx/sites/all/doc/cartillas/2015-2016/43-discriminacion-dh.pdf

Comisión Nacional para el Desarrollo de los Pueblos Indígenas. (2006). *Percepción de la imagen del indígena en México. Diagnóstico cualitativo y cuantitativo*. CDI. https://www.inpi.gob.mx/2021/dmdocuments/percepcion_imagen_indigena_mexico.pdf

Consejo Estatal de Población. (2019). *Envejecimiento Demográfico*. Gobierno del Estado de México. https://coespo.edomex.gob.mx/sites/coespo.edomex.gob.mx/files/files/2019/ENVEJECIMIENTO%20demografico.pdf

Consejo Estatal para el Desarrollo Integral de los Pueblos Indígenas del Estado de México. (2023). *Localización*. CEDIPIEM. https://cedipiem.edomex.gob.mx/localizacion

Consejo Nacional de Población. (2019). *Colección. Proyecciones de la población de México y las entidades federativas 2016-2050*. Secretaría de Gobernación. https://www.gob.mx/cms/uploads/attachment/file/487366/33_RMEX.pdf

Consejo Nacional para Prevenir la Discriminación. (2008). El trato social hacia las mujeres indígenas que ejercen trabajo doméstico en zonas urbanas. *Documento de trabajo No. E-08-2008*. CONAPRED. http://cedoc.inmujeres.gob.mx/lgamvlv/CONAPRED/conapred07.pdf

Consejo Nacional para Prevenir la Discriminación. (2011). *30 de Marzo Día de las trabajadoras del hogar*. CONAPRED.

Consejo Nacional para Prevenir la Discriminación. (2015). *30 de marzo. Día internacional de las trabajadoras del hogar*. CONAPRED.

Consejo Nacional para Prevenir la Discriminación. (2018). *Informe Anual de Actividades y Ejercicio Presupuestal 2017*. CONAPRED. https://www.conapred.org.mx/wp-content/uploads/2023/05/InformeAnual2017.pdf

Consejo Nacional para Prevenir la Discriminación. (2019). *Informe Anual de Actividades y Ejercicio Presupuestal 201.8*. CONAPRED. https://www.conapred.org.mx/wp-content/uploads/2023/05/InfomeAnual2018.pdf

Consejo Nacional para Prevenir la Discriminación. (2020). *Informe Anual de Actividades y Ejercicio Presupuestal 2019*. CONAPRED. https://www.conapred.org.mx/wp-content/uploads/2023/05/InformeAnual2020.pdf

Consejo Nacional para Prevenir la Discriminación. (2021). *Informe Anual de Actividades y Ejercicio Presupuestal 2020*. CONAPRED. https://www.conapred.org.mx/wp-content/uploads/2023/05/InformeAnual2021.pdf

Consejo Nacional para Prevenir la Discriminación. (2022). *Informe Anual de Actividades y Ejercicio Presupuestal 2021*. CONAPRED. https://www.conapred.org.mx/wp-content/uploads/2023/05/InformeAnual2021.pdf

Consejo Nacional para Prevenir la Discriminación. (2023a). *Discriminación en contra de la población y pueblos indígenas*. Ficha Temática. CONAPRED. http://www.conapred.org.mx/wp-content/uploads/2024/02/FT_Pindigenas_Noviembre2023.pdf

Consejo Nacional para Prevenir la Discriminación. (2023b). *Informe Anual de Actividades y Ejercicio Presupuestal 2022*. CONAPRED. https://www.conapred.org.mx/wp-content/uploads/2023/06/InformeAnual2022.pdf

Consejo Nacional para Prevenir la Discriminación. (n.d.a). *Informes anuales del CONAPRED 2004-2020*. CONAPRED. https://www.conapred.org.mx/transparencia/planes-programas-e-informes/informes-anuales-del-conapred-2004-2020/

Consejo Nacional para Prevenir la Discriminación. (n.d.b). *Programa Nacional para la Igualdad y No Discriminación (PRONAIND)*. CONAPRED. https://www.conapred.org.mx/pronaind/

Consejo para Prevenir y Eliminar la Discriminación de la Ciudad de México. (2021). *COPRED llama a visibilizar las brechas de género persistentes en el Día Internacional de la Mujer Indígena*. COPRED. https://www.copred.cdmx.gob.mx/comunicacion/nota/copred-llama-visibilizar-las-brechas-de-genero-persistentes-en-el-dia-internacional-de-la-mujer-indigena

Consejo para Prevenir y Eliminar la Discriminación de la Ciudad de México, Centro Nacional para la Capacitación Profesional y Liderazgo de las Empleadas del Hogar A.C, Organización Internacional del Trabajo. (2021). *Informe sobre la situación de los derechos de las personas trabajadoras del hogar en la Ciudad de México*. COPRED, CACEH, OIT, Gobierno de la Ciudad de México. https://caceh.org.mx/wp-content/uploads/2022/02/informe-sobre-la-situacion-de-los-derechos-de-las-personas-trabajadoras-del-hogar-en-la-ciudad-de-mexico.pdf

Cortina, A. (2017). *Aporofobia, el rechazo al pobre. Un desafío para la democracia*. Paidós.

Cruz, R., Vargas, E., Hernández, A. y Rodríguez, O. (2017). Adolescentes que estudian y trabajan: factores sociodemográficos y contextuales. *Revista Mexicana de Sociología* 79(3), 571–604. https://revistamexicanadesociologia.unam.mx/index.php/rms/article/view/57679/51146

Cruz, R. y Zapata, R. (2013). Naturalización y vulnerabilidad de los inmigrantes mexicanos en Estados Unidos. En M. E. Anguiano y R. Cruz (Eds.), *Migraciones internacionales, crisis y vulnerabilidades. Perspectivas comparadas*. El Colegio de la Frontera Norte.

De Lama, A. (2013). Discriminación múltiple. *Anuario de derecho civil*, 66(1), 271–320. https://revistas.mjusticia.gob.es/index.php/ADC/article/view/3715

Durin, S. (2017). *Yo trabajo en casa: trabajo del hogar de planta, género y etnicidad en Monterrey*. CIESAS – Publicaciones de la Casa Chata

El Colegio de la Frontera Norte. (2019). *Caravanas migrantes de Centroamérica*. El Colef. https://www.colef.mx/estemes/caravanas-migrantes-de-centroamericanos/

Escalante, Y. (2009). Derechos de los pueblos indígenas y discriminación étnica o racial. *Cuadernos de la igualdad*, (11). https://sindis.conapred.org.mx/investigaciones/derechos-de-los-pueblos-indigenas-y-discriminacion-etnica-o-racial/

Esparza, M. (2012). Empleo insuficiente y deterioro de las condiciones laborales en Zacatecas en los albores del nuevo siglo. *Paradigma económico. Revista de economía regional y sectorial*, 4(2), 61–84. https://paradigmaeconomico.uaemex.mx/article/view/4782/3187

Estrada, J.P., Ávila, M. J., y Martínez, M. L. (2022). La discriminación histórica a personas migrantes en tiempos de la pandemia de la COVID-19 en Coahuila, México. *Huellas de la Migración*, 7(13), 11–43. https://doi.org/10.36677/hmigracion.v7i13.16595

Francioli, S. & North, M. (2021). Youngism: The content, causes, and consequences of prejudices toward younger adults. *Journal of Experimental Psychology General*, 150(12), 1–22. https://doi.org/10.1037/xge0001064

Gandini, L., Fernández, A., Narváez, J.C., Rodríguez, L.H., Franco, M., Pilatowsky, E. y Rojas, R. (2021). *Documento de trabajo 1. Protección social de las personas refugiadas y solicitantes de la condición de refugio en México. Un análisis de oportunidades y capacidades institucionales*. Organización Internacional del Trabajo. https://www.ilo.org/wcmsp5/groups/public/---americas/---ro-lima/---ilo-mexico/documents/publication/wcms_838086.pdf

Garabito, G. (2018). Trabajo y juventudes universitarias en México: tendencias y complejidades. En A. Corica, A. Freytes y A. Miranda (Coords.), *Entre la educación y el trabajo: la construcción cotidiana de las desigualdades juveniles en América Latina* (pp. 93–108). CLACSO.

García, B. (2010). Inestabilidad laboral en México: el caso de los contratos de trabajo. *Estudios Demográficos y Urbanos*, 25(1), 73–101. https://estudiosdemograficosyurbanos.colmex.mx/index.php/edu/article/view/1368/1361

García, T. (2012). El estatus de extranjería en México. *Propuesta de reforma migratoria*, 45(133), 55–91. https://revistas.juridicas.unam.mx/index.php/derecho-comparado/article/view/4734/6265

Giacaglia, M. (2002). Hegemonía. Concepto clave para pensar la política. *Tópicos. Revista de Filosofía de Santa Fe*, (10), 151–159. https://doi.org/10.14409/topicos.v0i10.7430

Gobierno de México. (2020). *Jornada Nacional de Sana Distancia.* Gobierno de México. https://www.gob.mx/salud/hospitalgea/documentos/jornada-nacional-de-sana-distancia

Gobierno de México (2022). Jóvenes Construyendo el Futuro. *Programas para el bienestar.* Gobierno de México. https://programasparaelbienestar.gob.mx/jovenes-construyendo-el-futuro/

Gobierno de México (2025). *Pensión para el bienestar de las personas adultas mayores.* https://programasparaelbienestar.gob.mx/pension-bienestar-adultos-mayores/

Gobierno de México. (n.d.). *Data México. Estado de México. Economía.* Gobierno de México. https://www.economia.gob.mx/datamexico/es/profile/geo/mexico-em?redirect=true

Gobierno del Estado de México. (n.d.). *Indicadores económicos.* Secretaría de Desarrollo Económico. https://desarrolloeconomico.edomex.gob.mx/indicadores_economicos

Godelier, M. (1986). *La producción de los grandes hombres. Poder y dominación masculina entre los Baruya de Nueva Guinea.* Akal.

Goffman, E. (2006). *Estigma. La identidad deteriorada.* Amorrortu.

Goldsmith, M. (1981). Trabajo doméstico asalariado y desarrollo capitalista. *Ideas feministas de Nuestra América*: https://ideasfem.wordpress.com/textos/i/i17/

González, A. y Aikin, O. (2023). (In)Movilidad humana en México en contextos de vulnerabilidad, crisis regionales y políticas de cierre de fronteras. *Análisis Plural* (3), 1–17. https://doi.org/10.31391/ap.vi3.45

González, C., y Sánchez, J.J. (2003). Efectos de un programa cognitivo-conductual para mejorar la calidad de vida en adultos mayores. *Revista Mexicana de Psicología*, 20 (1), pp. 43–58.

González, L. (2012). Trabajo del hogar y desigualdad de género. En Consejo Nacional para Prevenir la Discriminación (Comp.), *Dos mundos bajo el mismo techo. Trabajo del hogar y no discriminación* (pp. 91–99). CONAPRED.

González, L. (2015). *Organización espacial y social de la cocina mazahua en San Antonio Pueblo Nuevo, San José del Rincón (1950–2013)* [Tesis de Maestría, El Colegio de Michoacán A.C]. Repositorio de El Colegio de Michoacán. https://colmich.repositorioinstitucional.mx/jspui/handle/1016/309

González, R. (2010). Calidad de vida en el adulto mayor. En *Envejecimiento Humano. Una visión transdisciplinaria* (pp. 365–377). Instituto de Geriatría. https://www.gob.mx/inger/documentos/envejecimiento-humano-una-vision-transdisciplinaria

Gracia, M. A y Horbath, J. E. (2019). Condiciones de vida y discriminación a indígenas en Mérida, Yucatán, México. *Estudios Sociológicos, 37*(110), 277–307. https://doi.org/10.24201/es.2019v37n110.1666

Gramsci, A. (1975). *Cuadernos de la cárcel* (tomo 1). Ediciones Era.

Guiamet, J. y Saccone, M. (2013). Entre la educación y el trabajo: experiencias formativas de jóvenes trabajadores. *Avá. Revista de Antropología,* 22, 229–248.

Guillén, J. C., Menéndez, F. G., y Moreira, T. K. (2019). Migración: Como fenómeno social vulnerable y salvaguarda de los derechos humanos. *Revista de Ciencias Sociales (Ve),* XXV(E-1), 281–294. https://doi.org/10.31876/rcs.v25i1.29619

Gutiérrez, L. (2012). Mujeres indígenas trabajadoras del hogar. *Revista de derechos humanos - dfensor,* 1, 19–23 https://www.corteidh.or.cr/tablas/r27855.pdf

Gutiérrez, L. (n.d.). Trabajadoras del hogar indígenas en la Ciudad de México. *Las trece semillas zapatistas. Conversaciones desde los pueblos originarios.* https://tzamtrecesemillas.org/sitio/trabajadoras-del-hogar-indigenas-en-la-ciudad-de-mexico/

Gutiérrez, N. (2015). ¿Es una ventaja ser indígena en México en el siglo XXI? En Se*r indígena en México. Raíces y derechos. Encuesta Nacional de Indígenas* (pp. 29–164). Instituto de Investigaciones Jurídicas, UNAM.

Gutiérrez, J.M., Romero, J., Arias, S.R., y Briones, X.F. (2020). Migración: Contexto, impacto y desafío. Una reflexión teórica. *Revista de Ciencias Sociales (Ve),* XXVI(2), 299–313. https://doi.org/10.31876/rcs.v26i2.32443

Guzmán, R. y Jiménez, M.L. (2015). La Interseccionalidad como instrumento analítico de interpelación en la violencia de género. *Oñati Socio-legal Series*, 5(2), 596–612. http://ssrn.com/abstract=2611644

Heatley, A. (2021). Jóvenes y desigualdad en México: ¿el derecho de piso de una sociedad adultocéntrica? *Intersticios sociales*, (21), 71–98. https://www.intersticiossociales.com/index.php/is/article/view/305/569

Hernández, I. (2025). Movilidades laborales internas y metropolitanas desde comunidades rurales de la región noroeste del Estado de México. En A. Jardón (Coord.), *Escenarios de las movilidades y migraciones contemporáneas en el Estado de México*. Universidad Autónoma del Estado de México. http://ri.uaemex.mx/handle/20.500.11799/141973

Hernández, I y Jardón, A.E. (2018). Dinámicas contemporáneas de las movilidades rurales hacia las zonas metropolitanas de Toluca y Valle de México. El caso de la región noreste del Estado de México. En N. Baca, Z. Ronzón, R. Romo, R.P. Román y M. Padrón (Coords.), *Migración y movilidades en el centro de México* (pp. 171–189). CONAPO, UAEM.

Human Rights Watch. (2024). *Informe mundial 2024. Migrantes y solicitantes de asilo*. HRW. https://www.hrw.org/es/world-report/2024/country-chapters/mexico

Instituto de Liderazgo Simone de Beauvoir. (2023). *Cartilla informativa sobre atención de quejas y asesoría legal para personas trabajadoras del hogar remuneradas*. ILSB. https://www.gob.mx/cms/uploads/attachment/file/878020/Cartilla_ILSB.pdf

Instituto Mexicano de la Juventud. (2017). *¿Qué es ser joven?* Gobierno de México. https://www.gob.mx/imjuve/articulos/que-es-ser-joven

Instituto Mexicano del Seguro Social. (n.d.). *Beneficios*. IMSS. https://www.imss.gob.mx/personas-trabajadoras-hogar/beneficios

Instituto Mexicano del Seguro Social. (2024, 07 de octubre). *Puestos de trabajo afiliados al Instituto Mexicano del Seguro Social* [Comunicado de prensa]. https://www.imss.gob.mx/prensa/archivo/202410/009

Instituto Nacional de Estadística y Geografía. (2000). *Censo de población y vivienda 2000*. INEGI. https://www.inegi.org.mx/programas/ccpv/2000/#tabulados

Instituto Nacional de Estadística y Geografía. (2010). *Censo de Población y Vivienda 2010*. INEGI. https://www.inegi.org.mx/programas/ccpv/2010/

Instituto Nacional de Estadística y Geografía. (2020a). *Censo de población y vivienda 2020*. INEGI. https://www.inegi.org.mx/programas/ccpv/2020/#tabulados

Instituto Nacional de Estadística y Geografía. (2020b). *Noticia. Encuesta Telefónica sobre COVID-19 y Mercado Laboral (ECOVID-ML) abril – julio de 2020*. INEGI. https://www.inegi.org.mx/app/saladeprensa/noticia/6207

Instituto Nacional de Estadística y Geografía. (2020c). *Estadísticas a propósito del Día Mundial de la Población (11 De Julio)*. INEGI. https://www.inegi.org.mx/contenidos/saladeprensa/aproposito/2020/Poblacion2020_Nal.pdf

Instituto Nacional de Estadística y Geografía. (2022a). *Encuesta Nacional sobre Discriminación (ENADIS) 2022*. INEGI. https://www.inegi.org.mx/programas/enadis/2022/

Instituto Nacional de Estadística y Geografía. (2022b). *Encuesta Nacional de Ocupación y Empleo, Nueva Edición. Primer trimestre 2022*. INEGI. https://www.inegi.org.mx/programas/enoe/15ymas/

Instituto Nacional de Estadística y Geografía. (2023). *Para Saber +. Encuesta Nacional sobre discriminación*. INEGI. https://www.inegi.org.mx/contenidos/programas/enadis/2022/doc/enadis2022_infografia.pdf

Instituto Nacional de Estadística y Geografía. (2024, 26 de marzo). *Estadísticas a propósito del Día Internacional de las Trabajadoras del Hogar* [Comunicado de prensa núm. 204/24]. https://www.inegi.org.mx/contenidos/saladeprensa/aproposito/2024/EAP_tdom.pdf

International Labour Organization. (1989). *Indigenous and Tribal Peoples Convention, 1989 (No. 169)*. ILO. https://normlex.ilo.org/dyn/nrmlx_en/f?p=NORMLEXPUB:55:0::NO::P55_TYPE%2CP55_LANG%2CP55_DOCUMENT%2CP55_NODE:REV%2Cen%2CC169%2C%2FDocument

International Labour Organization. (2007). *Newsletter 2007. The ILO and the indigenous and tribal people. Theme: Discrimination.* ILO. https://www.ilo.org/wcmsp5/groups/public/@ed_norm/@normes/documents/publication/wcms_100542.pdf

International Labour Organization. (2011). *Domestic Work Convention, 2011,* (189). ILO. https://normlex.ilo.org/dyn/nrmlx_en/f?p=NORMLEXPUB:12100:0::NO::p12100_ILO_CODE:C189

Jacobo, M. y Cárdenas, N. (2021). Back on your own: migración de retorno y la respuesta del gobierno federal en México. *Migraciones internacionales,* 11 https://doi.org/10.33679/rmi.v1i1.1731

Jardón, A., López, A. y Martínez, N. (en prensa). Educación en Derechos Humanos. Claves para contrarrestar la xenofobia hacia la población en movilidad humana. En L. Delgadillo (Coord.), *Debates, desafíos y propuestas sobre vulnerabilidad e inclusión social.* UAEM, UNESCO.

Kalish, R. (1996). *La vejez: perspectiva sobre el desarrollo humano.* Pirámide.

Kuhner, G. (2011). La violencia a las mujeres migrantes en tránsito por México. *Opinión y debate,* (6). https://corteidh.or.cr/tablas/r26820.pdf

Lama, A. (2013). *Discriminación múltiple* (tomo LXVI). ADC. https://revistas.mjusticia.gob.es/index.php/ADC/article/view/3715

López, O. P. (2017). *Empoderamiento de las mujeres mazahuas del Estado de México. El caso de las que se quedan y las que se van de San Pedro del Rosal, Atlacomulco, 1950–1960* [Tesis de Licenciatura, Universidad Autónoma del Estado de México]. http://hdl.handle.net/20.500.11799/67656

Majidi, N. (2020). Assuming reintegration, experiencing dislocation—returns from Europe to Afghanistan. *International Migration,* 59(2), 186–201. https://doi.org/10.1111/imig.12786

Medellín, C. (2024). Huehuetoca el camino al norte, ruta de la esperanza para muchas familias de migrantes. *La Silla Rota.* https://lasillarota.com/metropoli/2023/9/23/huehuetoca-el-camino-al-norte-ruta-de-la-esperanza-para-muchas-familias-de-migrantes-449086.html

Montiel, O., Flores, A., Ávila, E. y Sierra, S. (2021). "Tengo que sobrevivir": Relato de vida de tres jóvenes microemprendedores bajo COVID-19. *Telos* 23(1), 67–81. https://doi.org/10.36390/telos231.06

Montoya, J. (2004). Los retos demográficos en el Estado de México. *Papeles de Población,* nueva época, 10 (40), 25–29. https://rppoblacion.uaemex.mx/article/view/8754/7461

Naciones Unidas. (1991). *Principios de las Naciones Unidas en favor de las personas de edad. Resolución 46/91 de la Asamblea General de las Naciones Unidas del 16 de diciembre.* NU. https://www.un.org/development/desa/ageing/resources/international-year-of-older-persons-1999/principles/los-principios-de-las-naciones-unidas-en-favor-de-las-personas-de-edad.html

Naciones Unidas. (2001). *Conferencia Mundial contra el Racismo, la Discriminación Racial, la Xenofobia y las Formas Conexas de Intolerancia.* NU. https://www.un.org/es/conferences/racism/durban2001

Naciones Unidas. (2010). *Principios de las Naciones Unidas en favor de las personas de edad.* https://www.un.org/development/desa/ageing/resources/international-year-of-older-persons-1999/principles/los-principios-de-las-naciones-unidas-en-favor-de-las-personas-de-edad.html

Naciones Unidas. (2018). *La Agenda 2030 y los Objetivos de Desarrollo Sostenible: una oportunidad para América Latina y el Caribe (LC/G.2681-P/Rev.3).* NU.

Naciones Unidas. (2023). *Temas. Relator especial sobre el derecho a la salud.* NU. https://www.ohchr.org/es/special-procedures/sr-health

Naciones Unidas. (n.d.a). *La Declaración Universal de los Derechos Humanos.* NU. https://www.un.org/es/about-us/universal-declaration-of-human-rights

Naciones Unidas. (n.d.b). *Objetivos de Desarrollo Sostenible.* NU. https://www.un.org/sustainabledevelopment/es/objetivos-de-desarrollo-sostenible/

Naciones Unidas México. (2022). *Informe de resultados 2022. Trabajando en conjunto para cumplir la promesa de no dejar a nadie atrás. Oficina de Coordinación Residente del Sistema de Naciones Unidas en México.* NU. https://mexico.un.org/es/232966-informe-de-resultados-2022

Nuvaez, J. (2019). La discriminación laboral en razón del género y la edad en Colombia. *Revista Arbitrada Interdisciplinaria Koinonía,* 4(7), 308-320. https://doi.org/10.35381/r.k.v4i7.207

Observatorio de la Juventud en Iberoamérica. (2019). *Encuesta de jóvenes en México 2019.* OJI. https://oji.fundacion-sm.org/nuestros-estudios/encuesta-mexicana-de-la-juventud/

Oehmichen, C. (2005). *Identidad, género y relaciones interétnicas. Mazahuas en la Ciudad de México.* UNAM-Instituto de Investigaciones Antropológicas.

Ojarasca. (2005). Cancionero de ausencias. *Ojarasca. La Jornada*, 95. Jornada. https://www.jornada.com.mx/2005/03/21/oja95-mazahua.html

Organización de los Estados Americanos. (2015). *Convención Interamericana sobre la Protección de los Derechos Humanos de las Personas Mayores. Washington.* OAS. https://www.oas.org/es/sla/ddi/docs/tratados_multilaterales_interamericanos_a-70_derechos_humanos_personas_mayores.pdf

Organización de las Naciones Unidas para la Educación, la Ciencia y la Cultura. (n.d.). *La educación transforma vidas.* UNESCO. https://www.unesco.org/es/education

Organización Internacional del Trabajo. (2014). *Guía sobre las normas internacionales del trabajo.* OIT. https://www.ilo.org/wcmsp5/groups/public/---ed_norm/---normes/documents/publication/wcms_246945.pdf

Organización Internacional para las Migraciones. (2015, 14 de julio). *Comunicado Global. La OIM apoya campaña de información en México para proteger a los migrantes* [Comunicado de prensa]. https://www.iom.int/es/news/la-oim-apoya-campana-de-informacion-en-mexico-para-proteger-los-migrantes

Organización Mundial de la Salud. (2021). *Década del envejecimiento saludable: informe de referencia. Resumen.* OMS. https://iris.who.int/handle/10665/350938

ONU Mujeres. (2016). *La CEDAW, Convención sobre los Derechos de las Mujeres.* ONU Mujeres. https://mexico.unwomen.org/es/digiteca/publicaciones/2016/01/la-cedaw-convecion-derechos-de-las-mujeres

Organización Panamericana de la Salud. (2021). *Informe mundial sobre el edadismo.* OPS. https://iris.paho.org/handle/10665.2/55871.

Organización Panamericana de la Salud. (2023). *La Convención Interamericana sobre la Protección de los Derechos Humanos de las Personas Mayores como herramienta para promover la Década del Envejecimiento Saludable.* OPS. https://iris.paho.org/handle/10665.2/57353

OXFAM México. (2023). *El muro mexicano. Estudio de percepción sobre la migración en México.* OXFAM México. https://oxfammexico.org/wp-content/uploads/2023/08/EMM_Informe_completoR4.pdf

Passel, J., D'Vera, C. & González-Barrera, A. (2012). Net Migration from Mexico Falls to Zero-and Perhaps Less. Pew Hispanic Center. https://www.pewresearch.org/race-and-ethnicity/2012/04/23/net-migration-from-mexico-falls-to-zero-and-perhaps-less/

Pelletier, P. (2014). La "discriminación estructural" en la evolución jurisprudencial de la Corte Interamericana de Derechos Humanos. *Revista IIDH*, 60, 205–215. https://www.corteidh.or.cr/tablas/r34025.pdf

Pérez, G. y Oliver, P. (2020). *Migraciones, derechos humanos y comunicación intercultural en los ambientes de trabajo. Cartilla de sensibilización.* Organización Internacional para las Migraciones. https://www.r4v.info/es/document/cartilla-de-sensibilizacion-migraciones-derechos-humanos-y-comunicacion-intercultural-en

Pérez, M. y Aguilar, M. (2021). #LadyFrijoles: señalamiento, discriminación y estigma de migrantes centroamericanos a través de redes sociales en México. *Andamios*, 18(45), 223–243. https://doi.org/10.29092/uacm.v18i45.817

Poder Ejecutivo del Estado de México. (2024). *Decreto por el que se cambia la denominación de la Ley del Adulto Mayor del Estado de México, por la Ley de las Personas Adultas Mayores del Estado de México, y por el que se reforma y adicionan diversas disposiciones de la misma.* https://legislacion.edomex.gob.mx/sites/legislacion.edomex.gob.mx/files/files/pdf/gct/2024/julio/jul171/jul171a.pdf

Pujadas, J. (2000). El método biográfico y los géneros de la memoria. *Revista de Antropología Social*, 9, 127–158. https://revistas.ucm.es/index.php/RASO/article/view/RASO0000110127A

Quiñonez, C. y Rodríguez, S. (2015). La reforma laboral, la precarización del trabajo y el principio de estabilidad en el empleo. *Revista Latinoamericana de Derecho Social* 21, 179–201. https://doi.org/10.22201/iij.24487899e.2015.21.9768

Ramírez, T. (2025). Cambios y continuidades en los patrones migratorios y movilidades poblacionales en el Estado de México. En A. Jardón (Coord.), *Escenarios de las movilidades y migraciones contemporáneas en el Estado de México*. Universidad Autónoma del Estado de México. http://ri.uaemex.mx/handle/20.500.11799/141973

Ramos, M. (2016). *Reconocimiento, derechos humanos e intervención social. Migrantes en el noreste de México*. México. Universidad Autónoma de Nuevo León.

Raphael, R. (Coord.) (2012). *Reporte sobre la discriminación en México 2012. Introducción general*. CIDE, Consejo Nacional para Prevenir la Discriminación.

Real Academia Española. (2023). *Diccionario de la lengua española*. Recuperado el 20 de noviembre del 2023, de https://www.rae.es/

Rincón, G. (2004). Presentación. En J. Rodríguez Zepeda, *Qué es la discriminación y cómo combatirla* (pp. 5–6). CONAPRED.

Rincón, G. (2005). Rasgos y retos de la lucha contra la discriminación en México. *El Cotidiano*, 21(134), 7–11.

Rodríguez, F. y Rossel C. (Coords.) (2009). *Panorama de la vejez en Uruguay*. UCU-UNFPA.

Rodríguez, J. (2004). *Qué es la discriminación y cómo combatirla*. CONAPRED.

Rodríguez, J. (2018). Sensatez y sensibilidad: cómo construir una institución antidiscriminatoria en un país fragmentado. En Consejo Nacional para Prevenir la Discriminación (Coord.), *Por la igualdad somos mucho más que dos. 15 años de lucha contra la discriminación en México* (pp. 49–64). Secretaría de Gobernación, CONAPRED.

Rodríguez, J. (2023). *Una teoría de la discriminación*. Universidad Autónoma Metropolitana-Iztapalapa.

Rodríguez, V. (2019). La discriminación interseccional en el discurso Jurídico. *Nuevo Derecho*, 15(25), 70–87. https://revistas.iue.edu.co/index.php/nuevoderecho/article/view/1235/pdf

Ronzón, Z., Méndez, A. y Jardón, A. (2021). El derecho a los cuidados de las personas mayores, una necesidad del sistema de Salud en México. *La Situación Demográfica de México 2021*. CONAPO. https://www.gob.mx/conapo/documentos/la-situacion-demografica-de-mexico-2021

Salomé, L. (2017). La discriminación y algunos de sus calificativos: directa, indirecta, por indiferenciación, interseccional (o múltiple) y estructural. *Pensamiento constitucional*, 22(22), 255–290. https://revistas.pucp.edu.pe/index.php/pensamientoconstitucional/article/view/19948/19969

Salvarezza, L. (Comp.) (1998). *La Vejez. Una mirada gerontológica actual*. Editorial Paidós.

Sánchez, C. (2004). La diversidad cultural en la Ciudad de México. Autonomía de los pueblos originarios y los migrantes. En P. Yanes, V. Molina y O. González (Coords.), *Ciudad, pueblos indígenas y etnicidad* (pp. 57–87). Universidad de la Ciudad de México.

Sánchez, E. y Zedillo, R. (2022). La complejidad del fenómeno migratorio en México y sus desafíos. *Serie de Documento de Política Pública. Elementos para entender los retos de la migración* (pp. 3–33). PNUD América Latina y El Caribe. https://www.undp.org/es/latin-america/publicaciones/la-complejidad-del-fenomeno-migratorio-en-mexico-y-sus-desafios

Sánchez, P., González, M., Ayala, B., Gutiérrez, L., De la Torre, K., Ventura, B. y Hernández, F. (2004). Sobre la experiencia y el trabajo de las organizaciones indígenas en la ciudad de México. En P. Yanes, V. Molina y O. González (Coords.), *Ciudad, pueblos indígenas y etnicidad* (pp. 321–368). Universidad de la Ciudad de México.

Sanz, A. (2005). El método biográfico en investigación social: potencialidades y limitaciones de las fuentes orales y los documentos orales. *Asclepio*, LVII, 99–115. https://doi.org/10.3989/asclepio.2005.v57.i1.32

Schuster, L. & Majidi, N. (2013). What happens post-deportation? The experience of deported Afghans. *Migration Studies*, 1(2), 221–240. https://doi.org/10.1093/migration/mns011

Secretaría de Gobernación. (2003, 11 de junio). *DECRETO por medio del cual se expide la Ley Federal para Prevenir y Eliminar la Discriminación*. DOF: 11/06/2003. https://www.dof.gob.mx/nota_detalle.php?codigo=694195&fecha=11/06/2003#gsc.tab=0

Secretaría de Gobernación. (2011, 10 de junio). *DECRETO por el que se modifica la denominación del Capítulo I del Título Primero y reforma diversos artículos de la Constitución Política de los Estados Unidos Mexicanos*. DOF:10/06/2011. https://dof.gob.mx/nota_detalle.php?codigo=5194486&fecha=10/06/2011#gsc.tab=0

Secretaría de Gobernación. (2020, 17 de diciembre). *Un logro, la inclusión de personas trabajadoras del hogar y agrícolas en la lista de salarios mínimos: Conapred*. [Comunicado de prensa]. https://www.gob.mx/segob/prensa/un-logro-la-inclusion-de-personas-trabajadoras-del-hogar-y-agricolas-en-la-lista-de-salarios-minimos-conapred

Secretaría de Gobernación. (2021). *Programa Nacional para la Igualdad y No Discriminación 2021–2024*. CONAPRED. http://www.conapred.org.mx/wp-content/uploads/2023/05/PRONAIND_2021-2024.pdf

Secretaría del Trabajo y Previsión Social. (2021, 03 de julio). *Entrada en vigor del Convenio 189 de la Organización Internacional del Trabajo protege a las personas trabajadoras del hogar* [Comunicado de prensa]. https://www.gob.mx/stps/prensa/comunicado-conjunto-017-2021?idiom=es

Soberanes, J. (2022). La discriminación en las convocatorias laborales. *Revista latinoamericana de derecho social, 35*, 271–296. https://doi.org/10.22201/iij.24487899e.2022.35.17279

Solís, P. (2017). *Discriminación estructural y desigualdad social. Con casos ilustrativos para jóvenes indígenas, mujeres y personas con discapacidad*. SEGOB, CONAPRED y CEPAL.

Solís, P. y Dalle, P. (2019). La pesada mochila del origen de clase. Escolaridad y movilidad intergeneracional de clase en Argentina, Chile y México. *Revista Internacional de Sociología, 77*(1), 1–17. https://doi.org/10.3989/ris.2019.77.1.17.102

Solís, P., Krozer, A., Arroyo, C. y Güemez, B. (2019). *Discriminación étnico-racial en México: una taxonomía de las prácticas. Documento de Trabajo # 1. Proyecto sobre Discriminación Étnico Racial en México (PRODER)*. El Colegio de México https://discriminacion.colmex.mx/wp-content/uploads/2019/08/dt1.pdf

SOS Racisme y Institut de Drets Humans de Catalunya. (2019). *Discriminación racial, discriminación por origen nacional: el caso de las leyes de migración y/o extranjería*. IDHC. https://www.idhc.org/es/publicaciones/discriminacion-racial-discriminacion-por-origen-nacional-el-caso-de-las-leyes-de-migracion-y-o-extranjeria.php

Taylor, S. y Bogdan, R. (1994). *Introducción a los métodos cualitativos de investigación* (2ª ed.). Paidós.

Téllez, Y., Ruiz, L., Velázquez, M. y López, J. (2013). Presencia indígena, marginación y condición de ubicación geográfica. *La situación demográfica de México 2013*. http://www.conapo.gob.mx/work/models/CONAPO/Resource/1738/1/images/7_Presencia_indigena_marginacion_y_condicion_de_ubicacion_geografica.pdf

The Office of the High Commissioner for Human Rights. (n.d.a). *CEDAW in your daily life*. OHCHR. https://www.ohchr.org/es/treaty-bodies/cedaw/cedaw-your-daily-life

The Office of the High Commissioner for Human Rights (n.d.b). *International Convention on the Elimination of All Forms of Racial Discrimination*. OHCHR. https://www.ohchr.org/en/instruments-mechanisms/instruments/international-convention-elimination-all-forms-racial

Trejo, A. (2017). Crecimiento económico e industrialización en la Agenda 2030: perspectivas para México. *Revista Problemas del Desarrollo, 188*(48), 83–111. https://www.probdes.iiec.unam.mx/index.php/pde/article/view/56026/51495

Unidad de Política Migratoria, Registro e Identidad de Personas. (2022a). *Diagnóstico de la movilidad humana en el Estado de México*. UPMRIP https://portales.segob.gob.mx/work/models/PoliticaMigratoria/CPM/foros_regionales/estados/centro/info_diag_F_centro/diag_E domex.pdf

Unidad de Política Migratoria, Registro e Identidad de Personas. (2022b). *Informe ejecutivo 2022. Pacto Mundial para una migración Segura, Ordenada y Regular en México*. UPMRIP https://portales.segob.gob.mx/work/models/PoliticaMigratoria/Documentos/Informe_PMM_2022.pdf

Unidad de Política Migratoria, Registro e Identidad de Personas. (2022c). *Boletines Estadísticos. Personas en situación migratoria irregular.* UPMRIP. http://www.politicamigratoria.gob.mx/es/PoliticaMigratoria/CuadrosBOLETIN?Anual=2022&Secc=3

United Nations. (2022). *World population prospects summary of results* [Informe UN DESA/POP/2022/TR/ NO. 3]. Departamento de Asuntos Sociales y Económicos. https://www.un.org/development/desa/pd/content/World-Population-Prospects-2022

Urrutia, C. y Cuenca, R. (2018). *Las desigualdades laborales que enfrentan los jóvenes en Lima metropolitana.* Instituto de Estudios Peruanos.

Varela, A., Ruíz, V. y Pech, C. (2021). Racismo, migración y discriminación. El trabajo de la re/presentación. *Andamios,* 18(45), 9–20. https://doi.org/10.29092/uacm.v18i45.808

Vargas, E. y Cruz, R. (2014). Búsqueda de empleo entre jóvenes de acuerdo con su participación y protección laboral en México. *Papeles de Población,* 20(81), 213–245. https://rppoblacion.uaemex.mx/article/view/8352

Vargas, Y., Santana, C., Torres, E., y Gutiérrez, S. (2020). Comparación de marcos conceptuales de la teoría de la discriminación de J. Rodríguez Zepeda y Adela Cortina. *Sincronía,* (77), 450–462.

Villagómez, P. (2010). El envejecimiento demográfico en México: niveles, tendencias y reflexiones en torno a la población de adultos mayores. En *Envejecimiento Humano. Una visión transdisciplinaria* (pp. 305–314). Instituto de Geriatría.

Weller, J. (2007). La inserción laboral de los jóvenes: características, tensiones y desafíos. *Revista de la CEPAL,* 92, 61–82. https://repositorio.cepal.org/entities/publication/70f4571b-0978-43ac-8faf-0ea31395403a

Weller, J. (2009). *El fomento de la inserción laboral de grupos vulnerables. Consideraciones a partir de cinco estudios de caso nacionales.* CEPAL.

Wong R. y Aysa, L. (2001). Envejecimiento y salud en México: un enfoque integrado. *Estudios Demográficos y Urbanos*, 16(3), 519–544. https://doi.org/10.24201/edu.v16i3.1107

Zúñiga, E. y García, J. (2008). El envejecimiento demográfico en México. Principales tendencias y características. *La situación demográfica de México 2008.* https://www.gob.mx/conapo/documentos/situacion-demografica-de-mexico-2008

The manufacturer's authorised representative in the EU is Springer Nature Customer Service Centre GmbH, Europaplatz 3, 69115 Heidelberg, Germany. If you have any concerns regarding our products, please contact ProductSafety@springernature.com

Printed and bound by CPI Group (UK) Ltd, Croydon, CR0 4YY

26/03/2026

02078953-0017